Diseases of Horticultural Crops
Identification and Management

Diseases of Horticultural Crops
Identification and Management

Sanjeev Kumar
Assistant Professor/Scientist
Department of Plant Pathology
College of Agriculture
Jawaharlal Nehru Krishi Vishwa Vidyalaya
Jabalpur - 482004, M.P., India

NEW INDIA PUBLISHING AGENCY
New Delhi – 110 034

NEW INDIA PUBLISHING AGENCY
101, Vikas Surya Plaza, CU Block, LSC Market
Pitam Pura, New Delhi 110 034, India
Phone: + 91 (11)27 34 17 17 Fax: + 91(11) 27 34 16 16
Email: info@nipabooks.com
Web: www.nipabooks.com

Feedback at feedbacks@nipabooks.com

ISBN : 978-93-83305-64-3

Composed and Designed by NIPA

Dedicated
To
My Parents

Preface

This would be a long preface if I was to set down in detail the reasons why I attempted to write this book in present form. Suffice it to say that the text is based on course which I have been teaching to undergraduate students. The purpose is to provide the basic and emerging facts whereby the students may be introduced to the scientific foundation of this noble course called Diseases of Horticultural Crops. To a certain extent any course is a reflection of the teacher's own training and of his special interest. The author as a regular teacher for undergraduate courses, has experienced that student do not take books. They rely solely on class notes. One of the reason being the non availability of text books covering entire course content, the way they are delivered. In the present book, I have tried to arrange the topics and the contents in such a manner that students get an easy reading with due continuity and thus understanding. It is hoped that the book will prove useful for teachers, professionals, and students in the discipline of plant pathology and horticulture.

I do not claim originality in the preparation of this book and has taken help from a large number of books, journals, periodicals, bulletins, internet, etc. I humbly thank authors, editors, and publishers of the books, journals, etc. I am also indebted to my wife Dr. (Mrs) Archana Rani for inspiration and help and children Saumya and Adyan for bearing with me during the preparation of this book.

Besides my best effort some mistakes, factual or printing, might have inadvertently crept on for which we solicit sorry in anticipation and if pointed to us, shall certainly correct them in subsequent edition. I shall also welcome suggestions and healthy criticism for the improvement of this publication.

June,2014 Sanjeev Kumar

Jabalpur

Contents

Diseases of Plantation Crops

Diseases of Vegetable Crops

Diseases of Flower Crops

Diseases of Fruit Crops

CHAPTER - 1

Diseases of Mango and Their Management

SI.No	Major diseases	Causal organism
1.	Anthracnose	*Colletotrichum gloeosporioides*
2.	Powdery mildew	*Oidium mangiferae (Acrosporium mangiferae)*
3.	Mango malformation	*Fusarium moniliforme* var. *subglutinans*
4.	Red rust	*Cephaleurus mycoides*
5.	Grey Blight	*Pestalotia mangiferae*
6.	Sooty mould	*Capnodium mangiferae*
7.	Stem end rot	*Diplodia natalensis*
8.	Bacterial leaf spot	*Pseudomonas mangiferae- indicae*

1. ANTHRACNOSE

Colletotrichum gloeosporioides (Glomerella cingulata)

Economic Importance

- Anthracnose disease of mango in association with leaf spots, blossom blight, and wiher tip is the most common and wide spread disease in India.
- It commonly occurs in the states of Punjab, Uttar Pradesh, Bihar, Assam and Kerala.
- Losses due to fruit infection have been estimated from 2-39%.
- Bombay green was completely wiped out in Tarai region of Uttar Pradesh due to severe wither tip.

- Mature fruits are blemished hence fetch low price in the market

Symptoms

- Produces leaf spots, blossom blight, wither tip, twigs blight and fruit rot.
- Small blister like spots develop on the leaves and twigs. Young leaves wither and dry. Tender twigs wither and die back symptom appears.
- Affected branches ultimately dry up. Black spots appear on fruits.
- The fruit pulp becomes hard, crack and decay at ripening.
- Infected fruits drop

Pathogen

- Mycelium septate and slightly dark coloured.
- Hyphae accumulate below the host cuticle and develop fruiting bodies called acervuli.
- Conidia single celled, hyaline, small and elongated.
- Setae are common on twigs but not on fruits.

Disease Cycle

- **Primary Infection:** Mainly through previous season infected leaves, defoliated branches, mummified flowers and flower brackets.
- **Secondary Infection:** Mainly due to transfer of conidia through rain splash or wind driven rain water.
- Lesions on leaves and weed found to be important sources of blossom and fruit infection.

Favourable Conditions

- Optimum temperature -25°C
- Relative humidity -95-97%
- A relative humidity of more than 95% for 12 h is considered essential for infection and development of pathogen on mango fruits.

Management

- Spray *P. fluorescens* at 3 weeks interval commencing from October at 5g/kg on flower branches. 5-7 sprays one to be given on flowers and bunches.
- Spray Carbendazim (0.1%) or Chlorothalonil (0.2%) at 15 days intervals until harvest.
- Before storage, treat with hot water, (50-55°C) for 15 minutes or dip in Benomyl solution (500ppm) or Thiobendazole (1000 ppm) for 5 minutes.

2. POWDERY MILDEW

Oidium mangiferae (Acrosporum mangiferae)

Economic Importance

- The disease is worldwide in distribution.
- The disease is present in almost all the states of the country and causes heavy loss every year.
- The disease has been first reported by Ajrekar from Bombay in 1917.
- Dropping of unfertilized infected flowers and immature fruits leads to serious losses.
- The losses has been estimated from 5-20% depending upon the weather conditions.

Symptoms

- The characteristic symptom of the disease is the white superficial powdery fungal growth on leaves, stalk of panicles, flowers and young fruits.
- Generally the infection begins from the tip of the inflorescence and spread downwards.
- The severity of the disease reduces both number and size of the fruit.
- The affected flowers and fruits drop pre-maturely reducing the crop load considerably or might even prevent the fruit set.

Pathogen

- The pathogen is an ectoparasitic biotroph .
- The hyphae are superficial , septate, hyaline, and ramify over the host surface forming white dense coating .
- Conidiophores short, hyaline and conidia single celled -barrel shaped, produced in chain.
- Fungus is odium type.

Disease Cycle

- Survives as dormant mycelium in affected leaves.
- Secondary spread by air borne conidia.
- Spores blown by wind from infected areas readily adhere to hairy, unopened flowers near tip of the inflorescence and germinate in five to seven hours.
- Fungus grows rapidly during cloudy weather accompanied with heavy morning mist.
- Full bloom and fruit set to pea size stages were most susceptible stage.

Favourable Conditions

- Warm humid weather with cool nights, heavy morning dew, and cloudy conditions are the most favourable environmental conditions for disease development.
- Optimum temperature- 15-30^{0}C.

Management

- Resistant varieties: Neelum, Zardalu, Banglora, Torapari-khurd and Janardhan pasand
- Two sprays with wettable sulphur 0.3% once before the flowers open and 2nd after the fruit set.
- Dusting twice or thrice with fine sulphur will check the disease.
- Spraying with Karathane (0.1%) or Cosan (0.2%) or Benlate (0.2%) before flowering and after fruit set (peanut stage).

3. MANGO MALFORMATION

Fusarium moniliforme var. *subglutinans*

Economic Importance

- Malformation is a serious threat to mango growing area of the world.
- Of late, the extent of this malady has taken such a magnitude that the mango industry is being threatened of its extinction in India, particularly in northern parts.
- It was first recognized in 1891 by Maries from Darbhanga, Bihar.
- The disease is mainly restricted to northern India, where over 50% of the trees are affected.

Symptoms

- Symptoms can be two types: floral malformation and vegetative malformation.
- In vegetative malformation affect mango seedlings and young plants, there is development of excessive vegetative branches of limited growth in seedlings. They are swollen with short internodes forming bunches of various size and the top of the seedlings shows bunchy top appearance.
- In floral malformation-floral parts get malformed. The malformed flowers are bigger in size. The malformed bunches in place of normal flower and fruit become conspicuous.

Pathogen

- Physiological disorder, eriophyid mites and virus were earlier suspected to cause this disease.
- It is now considered to be caused by *Fusarium moniliforme var. subglutinans.*
- Microconidia are one or 2 celled, oval to fusiform and produced from polyphialides.
- Macroconidia are rarely produced.
- They are 2 -3 celled and falcate.
- Chlamydospores are not produced.

Disease Cycle

- Primary Infection: Through malformed inflorescence
- Secondary Infection: Either by air borne conidia or by conidia carried by eriophid mite *Aceria mangiferae.*

Favourable Conditions

- Optimum temperature- 10-15^{0}C.
- More disease in young than in old plants.
- About 91 percent incidence in 4-8 years aged plants.

Management

- Use of disease free planting material.
- Pruning of diseased parts along with basal 15-20 cm apparently healthy portions.
- Spray Captafol (0.2%) or Carbendazim (0.1%).
- Spray napthelene acetic acid (NAA) 200 ppm during first week of October.

4. RED-RUST

Cephaleurus mycoides

Economic Importance

- Red rust has been reported from India.
- The disease can cause reduction in photosynthetic activity and defoliation which results in lowering the vitality of the plants.

Symptoms

- Algae attacks foliage and young twigs.
- Rusty spots appear on leaves, initially as circular, slightly elevated, coalesce to form irregular spots.
- The spots mature, fall off and leave cream to white valvet texture on the surface of the leaves.
- Initially the spots are greenish-grey in color and valvety in texture, but later on the surface bears reddish-brown appearance.

Pathogen

- *Cephaleurus virescens* after a period of vegetative growth develops its reproductive structures.
- Sporangia formed directly on the thallus are sessile and thick walled with orange pigments. Hey are formed singly on the vegetative filaments.
- When the sporangia are riped the contents are converted into zoospores and liberated through an opening in the wall.
- The zoospores are orange in colour, ovoid and swim actively by means of cilia

Disease Cycle

- Primary Infection: By zoospores.
- Diseased leaves serve as primary inoculum.
- Rain water has been found to be a source of spread of the alga .
- Infection has been found to be increased during rainy season.
- Common in close plantations.

Environmental Conditions

- Optimum temperature-25-30^{0}C.
- Relative humidity-high (More than 85%).
- Frequent rain with high wind velocity.

Management

- Avoid close plantations.
- Spray Bordeaux mixture (1%) or Copperoxy chloride (0.3%).

5. GREY BLIGHT

Pestalotia mangiferae

Economic Importance

- Grey blight was reported in India by Vala *et al.,* (1985) and Verma *et al.* (1991).

- It is weak parasite capable of infecting injured tissues, and healthy fruits in contact with the diseased ones.

Symptoms

- Brown spots develop on the margin and at the tip of the leaf lamina.
- They increase in size and become dark brown.
- Black dots appear on the spots which are acervuli of the fungus.

Pathogen

- Acervuli seen as minute black dots on affected portion.
- Mycelium is colored and septate.
- Conidia five celled middle three cells are colored and the end cells are hyaline Slender 3-5 appendages are produced at the apex of the spore.

Disease Cycle

- Survive on mango leaves for over a year.
- Spreads through wind borne conidia.

Favourable Conditions

- Temperature - 20-25°C
- Relative humidity-high

Management

- Remove and destroy infected plant parts.
- Spraying Copper Oxychloride (0.25%), Mancozeb (0.25%) or Bordeaux mixture (1.0%).

6. SOOTY MOULD

Capnodium mangiferae

Economic Importance

- Sooty mould is a serious disease and is very common in orchards where honey dew or sugery substances secreting insects *viz.*, mango hopper, scales, coccids and mealy bugs are found.

- The damage by the fungi is not direct but indirect, as they interfere with the normal functioning of plant by cutting off the effective leaf area for photosynthesis.
- Disease is severe in old and dense orchards where light intensity is low.

Symptoms

- The disease is characterized by the presence of a black velvety thin membraneous covering on the leaf surface.
- In severe cases, the tree appears black due to heavy infestation of moulds on entire surface of leaves and twigs which affect the photosynthetic activity.
- Because of the production of masses of black spores, which adhere to the leaf surface due to sticky honey dew the foliage appears black, ugly and hence the name 'sooty mould' is given.
- During flowering time, if the fungus infects the blossom, fruit setting is affected and sometimes even small fruits may fall.

Pathogen

- The mycelium of the fungi is superficial and does not penetrate the host tissue.

Favourable Conditions

- Sugary substance secreted by the insects is stated to be a condition favorable for development of sooty mould.
- Incidence of insects on the shoot is directly associated with disease severity.
- High humidity proved congenial for growth of the fungi.

Management

- Management should be done for insects and sooty moulds simultaneously.

- Application of bleaching solution (0.05–0.10%) has been found to be most appropriate.
- After that spray starch solution (1kg Starch/Maida in 5 litres of water. Boiled and dilute to 20 liters).
- Starch dries and forms flake which are removed along with the fungus.

7. STEM END ROT

Diplodia natalensis

Economic Importance

- Stem end rot is an important post harvest disease of ripe mango fruits and was first noticed by Chakravarti and Srivastava in 1964.
- Four to six percent fruits are normally spoiled annually in the Indian markets.
- The disease affects the export of fruit adversely.

Symptoms

- The dark epicarp around the base of the pedicel.
- In the initial stage the affected area enlarges to form a circular, black patch. Under humid atmosphere extends rapidly and turns the whole fruit completely black within two or three days.
- The pulp becomes brown and somewhat softer.
- Dead twigs and bark of the trees, spread by rains

Pathogen

- The fungus produces brown to black, globose to sub globose, pyriform, erumpent pycnidia that are ostiolate.
- Two types of conidia are produced within a pycnidium.
- One is thick walled and bicelled with four to six longitudinal striations. The another is hyaline, thin walled and unicellular.

Disease Cycle

- The twigs and bark of mango tree harbour the pathogen.
- With the onset of rains fungal spores are released in the orchard which causes infection.
- The avenues of natural infection on uninjured mango fruits are exposed surface of the attached pedicle or the stem end scar when pedicel is removed.

Favourable Conditions

- The temperature in the range of 32-36°C and relative humidity between 84-100 percent is suitable for development of disease.

Management

- Prune and destroy infected twigs.
- Spray Carbendazim or Thiophanate Methyl(0.1%) or Chlorathalonil (0.2%) at 15 days interval during rainy season.

8. BACTERIAL LEAF SPOT

Pseudomonas mangiferae- indicae

Economic Importance

- The first description of bacterial leaf spot was given by Patel *et al.* in 1948 from Maharashtra.
- Earlier the disease was known to attack mango cvs of southern India. Now its occurrence has been reported in the north Indian varieties also.
- It causes 10-100% loss to the mango crop in the field as well as in storage.

Symptoms

- Symptoms are seen on leaves and fruits.
- On the leaves initially slightly water soaked yellowish, translucent irregular spots are formed towards leaf tip which enlarge soon and become dark brown with a yellow halo.

- These dark spots are limited by veins and become angular resulting in cankerous raised lesions.
- Under high humid conditions these spots fuse, leaves turn yellow and drop down prematurely.
- When fruits are attacked, water soaked lesion develop and turn dark brown to black.
- The affected fruits become yellow with deep cracks severely affected fruits are shed prematurely.

Disease Cycle

- Primary infection is through infected leaves, branches, and fruits.
- Secondary infection occurs by wind driven rain and rain splash.

Management

- Fortnightly spray of 1% B.M. or Agrimycin - 100 or Plantomycin (200ppm) or Streptomycin (200 ppm)
- Resistant variety: Bombay green

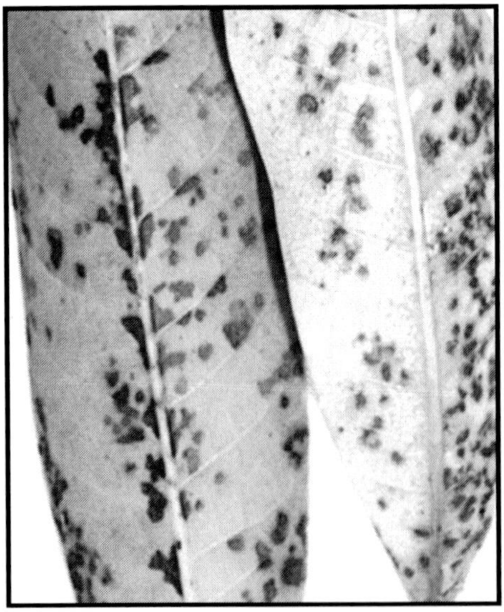

Anthracnose

Powdery mildew

Flower malformation

Sooty mould

Red rust

CHAPTER - 2

Diseases of Banana and Their Management

SI.No	Major diseases	Causal organism
1.	Panama wilt	*Fusarium oxysporum* f.sp *cubense*
2.	Moko	*Pseudomonas solanacearum / Burkholderia solanacearum*
3.	Sigatoka	*Mycosphaerella musicola*
4.	Anthracnose	*Gloeosporium gloeosporioides/Colletotrichum musae*
5.	Bunchy top	*Banana bunchy top virus*

1. PANAMA WILT

Fusarium oxysporum f. sp *cubense*

Economic Importance

- Panama disease is one of the commonest and most devastating diseases of banana worldwide.
- The disease was reported from Jamaica, Panama, Surinam, and other places at the end of previous century.
- The disease threatened the world banana trade at one stage in 1940s and 1950s which however, was saved by the development of the resistant 'Cavendish' cultivar.
- In India, the disease is a serious problem, and has been recorded from Andhra Pradesh, Tamil Nadu, Karnataka, Maharashtra and West Bengal.

- There is a report of development of a new virulent race of the pathogen, race-4, which threatens the banana industry once again.

Symptoms

- Symptoms are most clear on the plants that are at least five month old.
- Sudden wilting of the plant or individual leaves is a characteristic symptom of the disease.
- Yellowing of the lower most leaves starting from margin to midrib of the leaves.
- Yellowing extends upwards and finally heart leaf alone remains green for some time and it is also affected.
- The leaves break near the base and hang down around pseudostem.
- Discolored vascular strands varying from light yellow to dark brown are the distinguishing internal symptoms.
- Usually the discoloration manifests first in the outer or oldest leaf sheath and extends upto the pseudostem.

Pathogen

- Mycelium is septate, hyaline and branched.
- Fungus produces micro, macro conidia and also chlamydospores.
- Micro conidia - Single celled or rarely one septate hyaline elliptical or oval.
- Macro conidia - Sickle shaped hyaline, 3-5 septate and tapering at both ends.
- Chlamydospores - Thick walled, spherical to oval, hyaline to slightly yellowish in color.

Disease Cycle

- The pathogen is soil borne.
- The pathogens survives in the soil mainly as chlamydospores and invades the soft roots or rhizomes, very often through the injuries.
- The pathogen can also survive in the diseased rhizomes and other plant parts saprophytically for a long period.

- Primary infection always takes place through injured roots.
- Deep wounding exposing the xylem help in easy infection.
- When the germ tube of chlamydospores enters into the root, hyphae proceed internally along the root to the rhizome where they develop extensively in vascular tissues before passing up the vascular system into the pseudostem and the older leaf petioles.
- The secondary infection takes place by means of conidia.
- The disease can easily spread by contact of root system of adjacent healthy plants with conidia released by the diseased plants.
- Flood also help in the local dispersal of the secondary inoculums.

Favourable Conditions

- The disease is favoured by high soil moisture, light soils , bad drainage and the strong development of new roots.
- Optimum temperature-20-30^{0}C

Management

- Use of disease free suckers for planting.
- Dipping of suckers in Carbendazim (0.1%) solution before planting
- Corm injection of 3 ml of 2% Carbendazim injected in the corm by making a hole to a depth of 10 cm.
- Infected plant should be eradicated by weed killer.
- Resistant cultivar should be cultivated. Cavendish, Poovan, Moongil, Peyladen, Rajabale etc.
- Avoid ill drained soils, and prefer slightly alkaline soils (7-7.5 pH) for cultivation.
- Flood fallowing for 6 to 24 months or crop rotation with rice.
- Application of lime (1-2 kg/pit) to the infected pits after chopping of the plants parts.

2. MOKO OR BACTERIAL WILT

Pseudomonas solanacearum / Burkholderia solanacearum

Economic Importance

- First recorded in Guyana in 1840 in Moko plantain.
- In India, first reported from West Bengal in 1968.
- Moka is also very similar to the banana blood disease caused by race -2 of *R. solanacearum*
- In India, the fruit losses due to this disease, were upto 70%.

Symptoms

- Symptoms start on rapidly growing young plants
- The youngest three to four leaves turn pale green or yellow and collapse near the junction of lamina and petiole.
- Characteristic discoloration of vascular strands, wilting and blackening of suckers.
- Vascular discolouration (pale yellow to dark brown or bluish black) is concentrated near the centre of the pseudostem, becoming less apparent on the periphery.
- Greyish brown bacterial ooze is seen when the pseudostem of affected plant is cut transversely.
- A firm brown dry rot is found within fruits of infected plants (characteristic symptom).

Pathogen

- Moko disease is caused by race 2 of *Ralstonia solanacearum.*
- It is rod shaped, gram negative bacterium with one polar flagellum.

Disease Cycle

- The pathogen is soil borne.
- The pathogen survives in crop residues left in the field.
- Primary infection occurs when the bacteria enter the roots and rhizomes via wounds, caused by nematodes and farm implements.
- Secondary infections from plant to plants are caused by insects that transmit the bacterial oozes from infected male flowers to healthy flowers.

Favourable Conditions

- Warm and moist weather favours disease development.

Management

- Eradicate infected plant.
- Summer ploughing.
- Use of clean planting material.
- Fallowing and crop rotation is advisable.
- Disinfestation of tools with formaldehyde diluted with water in 1:3 ratio.
- Crop rotation (3 years rotation with sugarcane or rice) & providing good drainage.
- Fumigation of infected site with Methyl Bromide or Chloropicrin.
- Spray systemic insecticide to prevent transmission of disease to healthy plants.
- Biocontrol with *Pseudomonas fluorescens.*

3. SIGATOKA DISEASE

Mycosphaerella musicola (Cercospora musae)

Economic Importance

- First observed in Java in 1902.
- Epidemic in 1913 in plains of Sigatoka in Fiji Island.
- Two forms of Sigatoka leaf spots affect banana: yellow and black. Both diseases are of great importance, but black sigatoka is more virulent and difficult to control.
- It has become a threat to banana export industry.
- All commercial varieties are badly attacked and Gros Michel was virtually eliminated.

Symptoms

- Early symptoms appear on the young leaves.
- Small spindle shaped spots on foliage with greyish centre and yellowish halo running parallel to veins.

- Spots are mostly seen along the edge of the leaf with defined margin and possess dark brown to black margin. Spots coalesce and whole leaf blade dries up.
- On the upper surface of the spots, fructifications of the fungus appear as black specks.
- Unevenly and individual bananas appear undersized and their flesh develops a buff pinkish colour, and store poorly.

Pathogen

- Mycelium is hyaline, septate and branched.
- Conidia are elongated, narrow and multiseptate.
- Perithecia are dark brown to black and ostiolate.
- Asci are oblong and clavate.
- Ascospores are hyaline, two celled, obtuse to ellipsoid

Disease Cycle

- Survives on dry infected leaves on the field soil and primary infection takes place through ascospores in the infected plant debris.
- Secondary spread through wind borne conidia and ascospores.
- Infection takes place through stomata on the lower surface of young leaves.
- Surface moisture is necessary for release of both conidia and ascospores. Hence the disease is severe in moist weather.

Favourable Conditions

- High humidity, heavy dew and rainy weather with temperature above 21^0C.
- Soils with poor drainage and low fertility favour the disease incidence.
- Thick planting, presence of weeds and increased number of suckers in a mat promote disease development.

Management

- Removal and destruction of the affected leaves.

- Prevent water accumulation around the plant and go for periodical weeding.
- Spray Zineb (0.25%) or Copper oxy chloride (0.3%) suspended in mineral oil.
- Spray Chlorothalonil (0.2%) or Carbendazim or Thiophanate methyl (0.1%) or Mancozeb (0.25%) along with spreading agent. Wetting agent such as teepol or sandovit added at the rate of 1ml/lit of water.

4. ANTHRACNOSE

Gloeosporium gloeosporioides / Colletotrichum musae

Symptoms

- The skin at the distal ends of the fingers turn black, shrivels.
- The fungus produces masses of conidia which form a pinkish coat.
- The entire fruit and bunch is affected in severe cases. Sometimes main stalk of bunch diseased. The bunch becomes black and rotten.
- Acervuli produces cylindrical conidiophores, hyaline, septate, branched. Conidia hyaline, non-septate, oval to elliptical.

Pathogen

- Acervuli are usually rounded or sometimes elongated, erumpent.
- Conidiophores are cylindrical, tapered towards the apex, hyaline and septate.
- Conidia are hyaline, septate, oval to ellipitical in shape.

Disease Cycle

- The spread of the disease is by air borne conidia and numerous insects which frequently visit banana flowers also spread the disease.

Management

- Post harvest dipping of fruits in Carbendazim (400 ppm), or Benomyl (1000 ppm), or Aureofunginsol (100 ppm).

5. BUNCHY TOP

Banana bunchy top virus

Economic Importance

- The disease was first reported from Fiji in 1889 in Cavendish varieties.
- It also occurs in Australia, Sri Lanka and Egypt.
- In India, the disease is believed to have been introduced through infected suckers brought from Sri Lanka into the state of Kerala.
- The disease is covered by domestic quarantine regulations.
- It has now spread to almost all the banana growing areas in India.
- Losses were estimated to be Rs.4 crores every year and 100% loss occurs if infected suckers are planted.

Symptoms

- Subsequent leaves show the same symptoms and are dwarfed.
- Dark broken bands of green tissues on the veins, leaves and petioles.
- Plants are extremely stunted.
- Leaves are reduced in size, marginal chlorosis and curling.
- Leaves upright and become brittle.
- Many leaves are crowded at the top. Branches size will very small. If infected earlier no bunch will be produced.

Transmission

- The virus is not transmitted by mechanical inoculation.
- The chief insect vector is *Pentalonia nigronervosa.*
- The aphid acquires the virus in a feeding period of 24 hours and transmits the virus to healthy plants to an infection feeding period of one and a half years.
- The aphid retains the virus for a period of 13 days and the incubation period for the virus is a few hours to two days.

Perpetuation

- The virus perpetuates through infected plant suckers.

Management

- Adaptation of strict quarantine measures.
- Eradication of all infected suckers by spraying with kerosene or by injecting herbicide, 2, 4-D.
- Certified virus free suckers should be used for planting.
- Control vector by spraying Methyl demoton (1 ml/lit.) or Monocrotophos, (2 ml/lit.) or Phosphomidon (1 ml/lit.) or Injection of Monocrotophos 1 ml/plant (1 ml diluted in 4 ml). Infected plants are destroyed using 4ml of 2, 4, D (50g in 400 ml of water).

Panama wilt

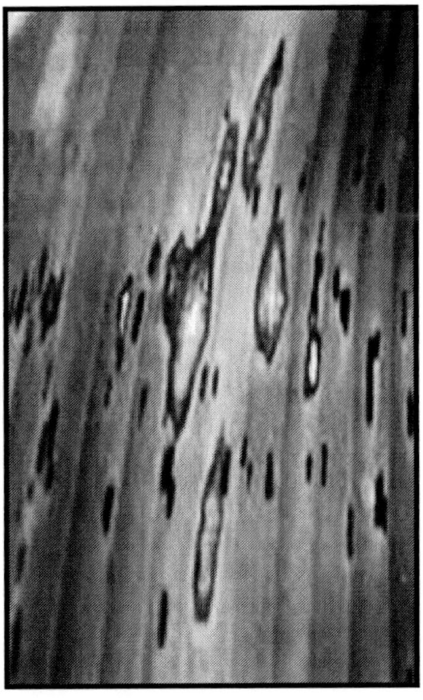

Sigatoka leaf spot

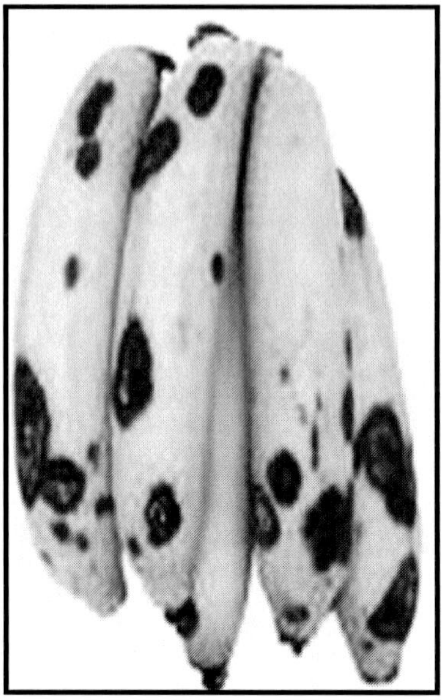

Anthracnose

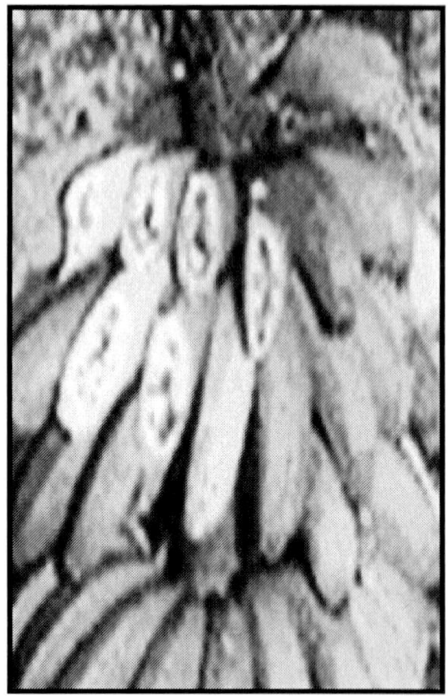

Bacterial wilt

Bunchy top

CHAPTER - 3

Diseases of Grapes and Their Management

SI.No	Major diseases	Causal organism
1	Downy mildew	*Plasmopara viticola*
2	Powdery mildew	*Uncinula necator*
3	Bird's Eye Spot/ Anthracnose	*Gloeosporium ampelophagum* (*Elsinoe ampelina*)

1. DOWNY MILDEW

Plasmopara viticola

Economic Importance

- Downy mildew is one of the most destructive diseases of grapevine.
- It occurs in most of the grapevine growing areas of the world.
- Economic loss is associated with cluster destruction and loss of vine foliage.
- Losses from 50 to 100 percent occur when conditions favour downy mildew.
- Since 1875, this disease caused heavy losses to wine industry in France.
- It led to discovery of Bordeaux mixture by Prof. Millardet in 1885.
- In India, the disease is quite common in Maharashtra and South India particularly in Karnataka and Andhra Pradesh.

Symptoms

- Symptoms appear on all aerial and tender parts of the vine.
- Irregular, yellowish, translucent spots on the upper surface of the leaves.
- Correspondingly on the lower surface, dirty white, powdery growth of fungus appears.
- Affected leaves become yellow and brown and gets dried due to necrosis.
- Premature defoliation.
- Dwarfing of tender shoots.
- Infected leaves, shoots and tendrils are covered by whitish growth of the fungus.
- White growth of fungus on berries which subsequently becomes leathery and shrivels. Infected berries turn hard, bluish green and then brown.
- Later infection of berries results in soft rot symptoms. Normally, the fully grown or maturing berries do not contact fresh infection as stomata turn non-functional.
- No cracking of the skin of the berries.

Pathogen

- The pathogen is biotroph.
- Mycelium is intercellular with spherical haustoria, coenocytic, thin walled and hyaline.
- Sporangiophores arise from hyphae in the sub stomatal spaces.
- It branched at right angle to the main axis and at regular intervals. Secondary branches arise from lower branches.
- The sporangia are thin walled, oval or lemon shaped.
- The sporangia germinate either by forming zoospores, or behave as conidia and germinate by germ tube.
- The zoospores are egg shaped, biflagellate and 7 - 9 micron meter. The oospores are thick walled.

- The oospores are thick walled with a rough epispore and are produced mostly in tissues adjacent to the midrib.

Disease Cycle

- The perpetuation of fungus from season to season is brought about by oospores present in the infected leaves, shoots and berries and also as dormant mycelium in infected twigs.
- Oospores present in soil germinate on maturity in favourable conditions and produces sporangia which, in turn, give rise to zoospores.
- The zoospores are transported by wind or water to the wet lower leaves near the soil line and causes primary infection.
- Secondary infection, for the first time during growing season, takes place by the sporangia produced on primary lesions.
- These sporangia are detached by movement of leaves and are wind borne to new infection sites on leaves and other parts to cause secondary infection.
- Sporangiophore and sporangial production takes place only in nights.
- Sporangia or zoospores are also distributed by rain .

Favourable conditions

- Optimum temperature : 20-22°C.
- Relative humidity : 80-100 per cent.

Management

- Collect and burn fallen leaves and twigs.
- Sanitation of the orchard.
- Vine should be kept high above ground to allow circulation of air by proper spacing.
- Pruning (April- May & September-October) and burning of infected twigs.
- Grow resistant varieties like Amber Queen, Cardinal, Champa, Champion, Dogridge and Red Sultana
- The disease can be effectively managed by giving 3-5 prophylactic sprays with 0.2% Fosetyl -Al (Aliette) or

Metalaxyl + Mancozeb (0.3 to 0.4%) or Azoxystrobin or Dimethomorph

2. POWDERY MILDEW

Uncinula necator

Economic Importance

- The disease is world wide in occurence and occurs commonly in Europe, USA, Africa and Australia in mild to severe form.
- In India,the disease appear in almost epidemic form in all the vineyards when the conditions are favourable for its development.
- The Losses in yield of fruits due to this disease may be upto 40-60%.
- In addition to loss of yield, infected berries tend to be higher in acid content than healthy fruits and are unsuitable for wine making.
- Powdery mildew disease of grapevines is quite serious disease of the crop in comparison to downy mildew of grapes.

Symptoms

- Powdery growth mostly on the upper surface of the leaves.
- Malformation and discolouration of affected leaves.
- Discolouration of stem to dark brown.
- Floral infection results in shedding of flowers and poor fruit set.
- Early berry infection results in shedding of affected berries.
- Powdery growth is visible on older berries and the infection results in the cracking of skin of the berries.

Pathogen

- White growth consists of mycelium, conidiophores and conidia.
- Mycelium is external, septate and hyaline.

- Conidiophores are short and arise from external mycelium. Conidia are produced in chain. They are single celled, hyaline and barrel shaped.
- The fungus is oidium type.
- When the mating types are present, cleistothecia can form on all infected tissues during later part of the growing season.
- Mature cleistothecia are black, depressed, contains 4-8 asci.
- 4-6 oval ascospores occur within each ascus.

Disease Cycle

- The cleistothecia and hyphae in diseased buds are the perennating structures of the disease.
- Since cleistothecia are not found in India, these are the hyphae in disease buds that serve the prupose.
- The primary infection in a new growing season is established after infection from ascospores (where cleistotheica are formed) or from the perennating mycelium in the host buds.
- When the bud open in the next season these resting structures produce abundance of conidia which are wind borne and serve as secondary inoculum.

Favourable Conditions

- Cool dry weather.
- Optimum temperature in the range of 20-25 ^{0}C with enough R.H. favour disease incidence (November and December).

Management

- Clean cultivation of vines or removal and destruction of all diseased parts.
- Dustings of vines with 300 mesh Sulphur (1st when new shoots are 2 weeks old, 2nd prior to blossoming, 3rd when the fruits are half ripe).
- Spray Wettable Sulphur (0.3%) or Karathane or Calixin (0.1%) sprayed at 15 days interval starting from last week of December to 1st week of March.
- Grow resistant varieties like Flame Muscat, Chholth Red, Chholth White, Skibba Red, Skibba White, etc.

3. ANTHRACNOSE /BIRD'S EYE SPOT

Gloeosporium ampelophagum (*Elsinoe ampelina*)

Economic Importance

- Anthracnose has been found to occur in all the grape growing countries.
- In India, anthracnose has become a potential threat to grape cultivation in North India.
- It appears every year in these areas and reduces the quality and quantity of the crop apart from making the vines weak.

Symptoms

- Symptoms visible on leaves, stem, tendrils and berries.
- Young shoots and fruits are more susceptible than leaves.
- Circular, greyish black spots or red spots with yellow halo appear.
- Later the centre of the spot becomes grey, sunken and fall off resulting in a symptom called 'shot hole'.
- Black, sunken lesions appear on young shoots.
- Cankerous lesions on older shoots. Girdling and death of shoots occur.
- Infection on the stalk of bunches and berries result in the shedding of bunches and berries, respectively.
- Sunken spots with ashy grey centre and dark margin on fruits (Birds eye symptom).
- Mummification and shedding of berries.

Pathogen

- The anthracnose is caused by *Gloeosporium ampelophagum*
- The fungus produces acervuli on the upper surface of the lesions.
- Acervuli contain short, clindrical conidiophores bearing small, ovoid, hyaline, conidia.
- Asci are globular and contain eight ascospores.

Disease Cycle

- The pathogen overwinters as dormant mycelium and sclerotia in the cankers which remain deep seated on the branches.
- They are able to cause fresh infections in spring on young shoots by producing conidia under favourable conditions.
- Conidia can germinate and infect with incubation period varying from 3-14 days.

Favourable Conditions

- Humid and wet weather conditions are highly favourable for development of disease.
- Optimum temperature-24-26^{0}C.
- Leaf wetness period-3 hrs.

Management

- Removal of infected twigs.
- Spray of Copper oxychloride (0.2%) or Mancozeb (0.2%).

Downy mildew

Powdery mildew

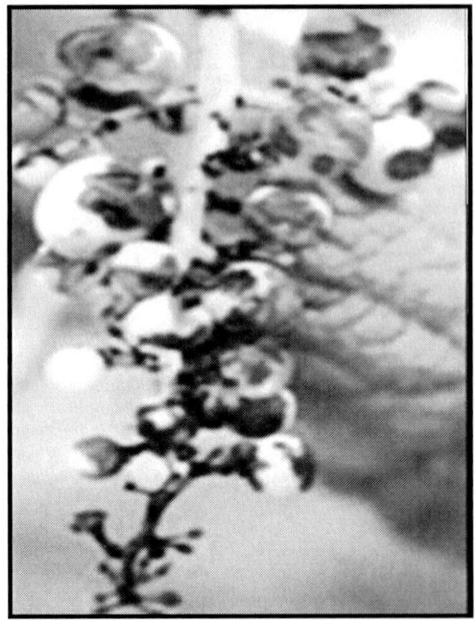

Anthracnose

CHAPTER - 4

Diseases of Citrus and Their Management

SI.No	Major diseases	Causal organism
1.	Gummosis	*Phytophthora parasitica, P. palmivora,P. citrophthora*
2.	Canker	*Xanthomonas campestris pv citri*
3.	Scab/Verucosis	*Elsinoe fawcett*
4.	Tristeza or quick decline	*Citrus tristeza virus*
5.	Greening	*Liberobactor asiaticum*
6.	Exocortis or Scaly butt	Viroid

1. GUMMOSIS

Phytophthora parasitica, P. palmivora, P. citrophthora

Economic Importance

- Gummosis, also known as brown rot gummosis, collar rot, trunk rot or root rot is present in almost all the countries where citrus is cultivated.
- In India, the disease occurs sporadically in states like Punjab Gujrat, Andhra Pradesh, Karnataka and Assam but prevalent in Maharashtra.
- Sweet varieties of citrus are more susceptible than sour ones and if the disease is not checked in time the entire tree may be destroyed.

Symptoms

- First symptoms are dark staining of bark which progresses into the wood.
- Bark at the base is destroyed resulting in girdling and finally death of the tree.
- Bark in such parts dries, shrinks and cracks and shreds in lengthwise vertical strips.
- Later profuse exudation of gum from the bark of the trunk(characteristic symptom).
- The infection spreads both upward to the leaves and even fruits and downward to the roots causing fibrous root rot.
- The fibrous root rot destroys a major portion of the root system.
- However, the trunck and root infections may usually result in ultimate drying of the tree.

Pathogen

- *P. palmivora* is the main cause of the disease in southern parts of our country although *P. parasitica* is also common in Karnataka *P. parasitica* is found to cause the disease in Assam

Character	*P. Palmivora*	*P. Parasitica*
Hyphae	Intercellular with haustoria and often swollen at regular intervals	Tough and irrregular
Sporangiophores	Simple or branched	Slender, irregular or sympodically branched
Sporangia	Obpyriform and always terminal	Lateral or intercalary, broadly ovoid, obpyriform to spherical,papillate
Oospores	Spherical, thickwalled and produce secondary sporangium on germination	Apleurotic

Favourable Conditions

- Prolonged contact of trunk with water as in flood irrigation; water logged areas and heavy soils.

- Optimum temperature for disease development -25-28^{0}C
- Soil pH-5.4-7.5

Disease Cycle

- The disease is soil borne with oospores and chalymydospores serving as perennating structures.
- Primary infection takes place on the bark near the soil line producing water soaked spots on the trunk and the crown roots.
- Oospore and chlamydospore perennating in soil on return of faovuable conditions germinate directly by germ tube or indirectly by producing sporangia, which liberate biflagellate zoospores.
- The fungus survives in the cankerous spots as dormant mycelium, and becomes active when environmental conditions are favourable.
- Sporangia and zoospores are produced as a result of primary infection serve as secondary inoculum.
- They are spread by wind, rain drop splashes, irrigation water, and even insects.

Management

- Selection of proper site with adequate drainage.
- Provision of an inner ring about 45 cm around the tree trunk to prevent moist soil.
- Avoid irrigation water from coming in direct contact with the trunk.
- Avoid injuries to crown roots or base of stem during cultural operations.
- Use resistant sour orange rootstocks for propagating varieties
- Painting Bordeaux paste or with $ZnSO_4$, $CuSO_4$, lime (5:1:4) to a height of about 60 cm above the ground level at least once a year.
- Scrape the diseased portion with a sharp knife.
- Protect the cut surface with Bordeaux paste followed by spraying of 0.3% Fosetyl-AL reduces the spread.

- Soil drenching with 0.2% Metalaxyl and 5% *Trichoderma viride* commercial formulation is also effective.

2. CANKER

Xanthomonas campestris pv *citri*

Economic Importance

- The disease was first noticed in Japan in 1904
- In India, citrus canker is endemic and occurs in all the citrus growing areas.
- It is an extremely persistent disease. Once established , it is very difficult to eradicate
- Acid lime is highly susceptible to canker.
- Severe foliage infection often causes defoliation, leaving only the bare twigs leading to almost complete loss.

Symptom

- The disease appears as raised , rough, corky brownish pustules with characteristic halo on leaves.
- Canker lesions on the fruit do not possess the yellow halo as on leaves.
- Several lesions on fruit may coalesce to form a patch. The crater-like appearance is more marked on fruits than on leaves.
- On fruits, canker lesions reduce market value.
- The plants are also stunted and fruit yields are reduced considerably.

Pathogen

- The pathogen is Gram negative, non spore forming, aerobic bacteria.
- It is rod shaped, forms chains and capsules and is motile by one polar flagellum.
- Three distinct forms (A, B and C) of the bacterium have been reported to exist on the basis of geographical distribution and host range.

Favourable Conditions

- Free moisture for 20 minutes.
- Favourable temperature- 20-30°C.

Disease Cycle

- The bacteria over winters on cankered lesions on the leaves twigs and fruits.
- Citrus leaf miners (*Phyllocnistis citrella*) helps in the dissemination of the pathogen.
- The bacterial cells start multiplying inside the host tissue during the onset of spring and ooze out in large numbers.
- Wind accompanying rain also causes infection as the thorny variety of citurs get wounded by the thorns moving in the wind , opening up wounds for the entry of the pathogen.
- Rain water also splashes the bacteria, which becomes wind borne and spread from place to place.

Management

- Prune badly infected twigs before the onset of monsoon.
- Periodical spraying of bactericides along with an insecticide.
- Use canker tolerant varieties.
- Select seedlings free from canker for planting in main field.
- Spray Streptocycline (1g) + Copper oxy chloride (30g) in 10 litres of water at fortnightly intervals for effective management of canker in citrus nurseries.
- Three sprayings of Streptocycline (1g) + Copper oxy chloride (30g) in 10 litres of water at an interval of 20 days during rainy season is effective for leaf canker.
- Fruit infection can be prevented by two sprays of Streptocycline (1g) + Copper oxy chloride (30g) in 10 litres of water at marble stage followed by another spray 30 days later.
- Control leaf miner when young flush is produced.

3. SCAB/VERUCOSIS

Elsinoe fawcett

Economic Importance

- Scab is one of the common diseases of sour oranges, rough lemon, tangelos and Pummelo.
- The market value of the fruits is considerably reduced due to this disease.
- In India, this disease is reported from Bengal, Punjab, Assam, Karnataka, Uttar Pradesh and Andhra Pradesh.
- The sweet oranges are mostly resistant to the disease.

Symptoms

- The disease is first noticed by the irregular, corky spots on the leaves, twigs and fruits.
- The spot may be a few mm upto one cm in diameter.
- They become warty, erupted with cracks and scabby, hence the name.
- On the fruits the warty growth are surrounded by yellowish or chlorotic haloes.
- The color of the spot turns from yellow to orange -brown and finally black.
- Several spots run together to form large patches of corky outgrowths.
- The scab may be a few or many on each fruits and the market value of the fruits is considerably reduced because of the infection.

Pathogen

- Ascostroma are simple, innate, intra or sub epidermal, partially erumpent at maturity, small pulvinate to crustose. Asci are ovoid.
- Ascospores are 1-3 septate oblong to elliptical and hyaline to yellowing conidia are produced in acervuli.
- Conidia are hyaline, oblong, elliptical with two minute droplets of their ends.

Favourable Conditions

- The fungus infects tissues only when the surface is wet and prefers temperature of 16 to 23^0C.
- The disease is a problem under low temperature and high humidity conditions.

Disease cycle

- The fungus produces mostly the conidial stage *Sphaceloma fawcetti*, on the host. The perfect stage (*E. fawcetti*) is produced *in vitro*.
- The fungus perpetuates and survives the off-season as ascospores.
- The secondary spread in the field is through air borne conidia.

Management

- The diseased leaves, twigs and fruits should be collected and destroyed.
- Spray with 0.3% Copperoxychloride or 1.0% Bordeaux mixture or 0.2% Difolaton.

4. TRISTEZA OR QUICK DECLINE

Citrus tristeza virus (CTV)

Economic Importance

- The name "Tristeza" was suggested to describe the sad appearance of the diseased citrus trees.
- This disease was first reported in *Citrus aurantifolia* and *C. sinensis* from Italy and Florida in the U.S.A.
- In India, tristeza is present in Andhra Pradesh, Tamil Nadu, Karnataka, Madhya Pradesh, Maharashtra, Punjab, Bihar, West Bengal and Sikkim.
- It affects all kinds of citrus plants but primarily orange, grapefruit and lime.
- Kagzi lime, grape fruits and Orlando tangelo are indicator plants for CTV detection.

Symptoms

- The infected trees usually show general decline.
- Leaves becomes dull and chlorotic and twigs start drying from tip downwards.
- In sweet orange or mandarin on susceptible root stocks, leaves develop deficiency symptoms and abscise.
- Roots decay, twigs die back, fruit set diminishes; only skeleton remains.
- Fine pitting of inner face of bark of sour orange stock.
- Acid lime leaves show large number of vein flecks (elongated translucent area).
- Tree stunted and dies, yield very much reduced. Fruits are small in size.

Pathogen

- Tristeza is a highly flexuous filamentous virus 2000 nm long and 10-12 nm in width.
- The virus is restricted to only phloem tissue and is a member of Closterovirus group.
- CTV has ss +ve sense RNA of 20 K nucleotides as its genome.

Mode of spread

- Transmission of virus occurs through the use of infected bud wood aphid vector like,*Toxoptera citricida.*

Management

- Strict quarantine measures to be enforced.
- Use certified budwood free of CTV.
- Remove all diseased trees as and when the disease is noticed.
- Fresh plantings to be taken with virus free materials on tolerant rootstocks.
- For the Punjab region, Jattikhatti, Cleopatra mandarin and sweet orange are recommended as resistant root stocks.

- For the Andhra Pradesh, Maharashtra and Karnataka, Rangapur lime is recommended as a root stock resistant to Tristeza.
- Periodic sprays of insecticides like Monocrotophos 0.05 % reduces the secondary spread of the disease in the orchard.

5. GREENING

Liberobacter asiaticum (Phloem limited bacterium).

Economic Importance

- The greening disease of citrus is widespread and is present all over the country.
- The disease is known to occur in Andhra Pradesh, Karnataka, Punjab, Uttar Pradesh, Himachal Pradesh, Rajasthan, Maharashtra, Jammu, Bihar, Bengal and Sikkim.
- In Punjab, the incidence of greening is as high as 64.5%
- The disease affects almost all citrus varieties irrespective of root stock.
- In India, mosambi, sweet orange, kagzi lime and orlando tangelo are good indicator plants for greening.

Symptoms

- The disease initiates as interveinal chlorosis and leads to the complete yellowing of the leaf with few green islands.
- Secondary growth is upright, short with small leaves and multiple buds.
- Leaf symptom pattern indicates zinc or iron deficiency.
- The fruits also show conspicuous sun blotching with yellow discoloration.
- As the disease advances, the plant shows severe die back symptoms.

Pathogen

- The disease is considered to be caused by a pholem limited *fastidious greening bacterium. Candidatus liberobacter asiaticum.*

Mode of Spread

- The disease is transmitted through infected bud wood and citrus psylla, *Diaphorina citri*.
- The disease also transmits from citrus to periwinkle (*Catharanthus roseus*) through dodder.

Management

- Select certified disease-free seedlings.
- Control psyllids with insecticides like dimethoate.
- Use pathogen free bud wood for propagation.
- Raise virus free plants through shoot tip grafting.
- 500 ppm tetracycline spray though effective, requires fortnightly application.

6. EXOCORTIS OR SCALY BUTT

Viroid

Economic Importance

- Exocortis was first reported from Maharashtra in India.
- The disease was later reported from Delhi and Punjab.
- At present, the disease is posing serious threat whereever Trifoliates, Rangpur lime and Citrange are used as root stocks.

Symptoms

- Vertical cracking and scaling of bark in the entire root stock.
- Extreme stunting of plant.
- Affects only rangpur lime, trifoliate orange and citrange root stocks.

Pathogen

- It is free RTVA without protein coat.

Transmission

- Transmission normally occurs through infected bud wood and contaminated tools.
- Not through vector and seed.

Management

- Removal of unproductive trees and replantation of viroid free trees.
- Spray with any one of the systemic insecticide to control the aphid vector.
- Use virus-free certified budwood and tolerant stocks like rough lemon.
- Periodically wash budding knife with Disodium Phosphate solution.

Canker

Gummosis

Scab

Tristeza or quick decline

Greening

CHAPTER - 5

Diseases of Guava and Their Management

SI. No	Major diseases	Causal organism
1.	Wilt	*Fusaium oxysporum f.*sp. *psidii, F. solani*
2.	Anthracnose	*Colletotrichum gloeosporioides* (*Glomerella psidii)*
3.	Canker	*Pestalotia psidii*

1. WILT

Fusaium oxysporum f. sp. *psidii, F. solani, Macrophomina phaseolina, Cephalosporium* sp.

Economic Importance

- This disease is very destructive and is known to occur in whole of northern India
- In India, the disease was first reported in 1935 from the Babakkarpur area of Allahabad and subsequently reported from other places of UP, West Bengal and recently from Haryana, Punjab and Rajasthan.
- The prestigious belt of guava *i.e.,* Allahabad is severely affected by this disease.
- The susceptibility of guava plants to wilt is known to decrease with age.

Symptoms

- The symptoms are browning and wilting of leaves, discoloration of the stem and death of the branches along one side.
- Sometimes the infection girdles the entire stem and the whole plant wilts.
- The leaves fall off and plant may die in the course of a year or more.
- When an infected stem is cut open the tissues extending up to the cambium region display a dark color.
- The fungus is present in the stem tissues, especially at the basal portions near the ground level.

Pathogen

- Das Gupta and Rai (1947) and Prasad *et al.*, (1952) attributed wilt disease in UP to infection of *F oxysporum f.*sp. *psidii.*
- In West Bengal, Chattopadhyay and Bhattacharya (1968) attributed the wilt to *Fusarium solani* under high soil moisture conditions and to *Macrophomina phaseolina* in dry soils.

Disease Cycle

- The fungus survive on dead root bits of guava in adverse climatic conditions in summer, while in rainy and winter seasons they survive around roots.
- The pathogen (*F oxysporum f.* sp. *psidii*) penetrates either through the root piliferous layer of the guava seedlings or through opening caused by secondary roots.
- After penetration, the intracellular mycelium first establishes in the epidermal cells and then spreads into the cortical cells which are considerably damaged and filled with the mycelium.
- *Fusarium solani* enter the xylem vessels, grows inside and blocks them.
- *M.phaseoli* first invades the phloem and destroy it.

Favourable Conditions

- pH 6.0 is optimum for disease development. Both pH 4.0 and 8.0 reduces the disease.
- Disease is more in clay loam and sandy loam compared to heavy soil.
- Higher disease incidence in monsoon period.
- Disease appears from august and increases sharply during September-October.

Management

- Proper sanitation of orchard.
- Pruning to remove branches 30-40 cm from discolored conducting vessels in March.
- Drenching of soil with 0.2% Benomyl or Carbendazim, four times in a year.
- Spray twice with 0.05% Metasystox and 0.3% Zinc sulphate twice a year.
- Soil solariztion with transparent polythene sheet during summer months.
- Application of oil cakes like neem cake, mahua cake, kusum cake supplemented with urea. Apply 6kg neem cake + 2kg gypsum per plant.
- Judicious ammendments of N and Zn.
- Resistant variety: Apple guava, Banarsi, Nasik, White Guava No. 6299, Supreme and Lucknow-49.
- Select only healthy seedling for planting of orchards.
- *Aspergillus niger* strain AN 17, *Trichoderma viride, Trichoderma harzianum* and *Penicillium citrinum* can used as biocontrol agents.

2. ANTHRACNOSE

Colletotrichum gloeosporioides (Glomerella psidii)

Economic Importance

- The disease was first reported from Uttar Pradesh.

- The disease is serious problem in Uttar Pradesh, and now of common occurence in all the guava growing region of India.
- It causes die back, twig blight, wither tip or fruit spots.
- Eighty percent of guava plants with 40% infected fruits with lower carbohydrates, cruid fibre, ash, fat, protein, Ca, Fe and P have been recorded due to anthracnose infection.
- The disease appears mostly during the rainy season.

Symptoms

- Symptoms of this disease are observed on mature fruits on the tree.
- The characteristic symptoms consist of sunken, dark colored, necrotic lesions. Under humid conditions, the necrotic lesions become covered with pinkish spore masses.
- As the disease progresses, the small sunken lesions coalesce to form large necrotic patches affecting the flesh of the fruit.
- The infected area in the unripe fruits become corcky and often develops cracks in case of severe infection.
- On ripe fruit, the infection causes softening of tissue and lesions attains a diameter of 10 to 20 mm infection.

Disease Cycle

- Primary Infection: Incipient infection carried by fruits from the field.
- Secondary Infection: Conidia by wind and rain splash.

Favourable Conditions

- The disease is found to develop more rapidly at 30^0C and 96.1percent relative humidity in both ripe and unripe fruits.

Management

- Use resistant varieties- Apple guava.
- Monthly spray of Copperoxychloride (0.3%) closely followed by Mancozeb (0.25%).

3. CANKER

Pestalotia psidii

Economic Importance

- The disease is prevalent in Mysore, Thane, Dharwar and Himachal Pradesh.
- No information is available from other parts of the country except Lucknow.

Symptoms

- The disease generally occurs on green fruits and rarely on leaves.
- Appearance of minute, brown or rust colored, circular, necrotic areas on fruits which in advanced stage of infection tears open the epidermis in a circinate manner.
- The margin of lesion is elevated and as depressed area is noticeable inside.
- The crater like appearance is noticeable on fruits than on leaves.
- The canker is confined to a very shallow area and does not penetrates deep into the flesh of the fruit.
- In older cankers, fruiting bodies of white mycelium consisting of numerous spores are noticeable.
- The infected fruit remain underdeveloped, become hard, malformed and mummified and great in numbers.

Pathogen

- The dark black and circular pycnidia on fruits contain conidiophores and conidia.
- The conidia are typically 5-celled, clavate or elliptic- fusoid, erect, constricted at septa, ;3 medium cells are gettulate, highly brownish, the central cell being the thickest and greatly bulged, and other cells are hyaline; the apical conical or cylindrical cell grows out into 3 hyaline, slender, elongated appendages, the basal cell is obtuse, erect with a small pedicel.

Favourable Conditions

- The pathogen grows vigorously at temperatures between 15-30^{0}C.
- Best growth and sporulation are seen at 26^{0}C.
- Maximum disease occurs at 25-30^{0}C.

Management

- Resistant cultivars; Safeda and Apple color
- 3 or 4 spraying of 1 percent Bordeaux mixture or lime sulphur (0.25%) at 15 days interval.

CHAPTER - 6

Diseases of Papaya and Their Management

SI.No	Major diseases	Causal organism
1.	Stem rot / Foot rot	*Pythium aphanidermatum*
2.	Powdery mildew	*Oidium caricae*
3.	Anthracnose	*Colletotrichum gloesporiodes*
4.	Ring spot	*Papaya ring spot virus*
5.	Leaf curl	Nicotiana virus 10

1. STEM ROT / FOOT ROT/COLLAR ROT

Pythium aphanidermatum

Economic Importance

- Foot rot is one of the most serious diseases of papaya and is widespread in India, Srilanka, Hawaii, Cuba and South Africa.
- In India, the disease generally manifests during rainy season(end of June to August) and becomes the cause of heavy damage to papaya.
- When severity of the disease is high, it devastates the entire plantations within the same growing season and makes the soil unfit for replanting.

Symptoms

- Spongy, water soaked spot in the stem at the ground level which enlarge and griddle the stem.
- The diseased area turns brown or black and rot.
- Rotting generally spreads above and below on the stem down to the roots which also begin to rot.
- Terminal leaves turn yellow droop off.
- The entire plant topples over and dies.
- Common in 2-3 year old plants.

Pathogen

- Mycelium is colorless, slender, coenocytic, profuse branching and rapidly growing mostly intracellularly.
- Sporangia terminal or intercalary produce a vesicle at the end of hyphal branch.
- Zoospores biflagellate and reniform.

Disease Cycle

- The pathogen perennates in soil through its oospores occurring in infected plant debris or most commonly through mycelium.
- The perennating oospore or saprophytic mycelium occurring in soil serve as the source of primary infection.
- Germ tubes resulted in by the germination of oospores, or of zoospores or saprophytic mycelium when it comes in contact with the ground level part of the plant, perennate therein and cause primary infection.
- The zoospores produced as a result of primary infection get disseminated by water currents, reach the ground level parts of the healthy papaya plants, and cause infection thereupon.
- On the later part of growth phage the pathogen growing as a result of primary or secondary infection undergoes sexual course of its life and produces oospores.
- The latter perennate and serve as the source of primary infection for the next growing season.

Favourable Conditions

- Excessive rains and relatively high temperature (around 36^0C).
- Appears from June to August.
- Younger seedlings are more susceptible than older ones.
- Severity increases with intensity of rainfall.

Management

- Seedlings should be raised in well drained nursery area.
- Uproot the diseased seedlings and burn.
- Water logging in the field should be avoided.
- Affected pit should not be used for replanting.
- Seed treatment with Captan @ 4g/kg seed or Chlorothalonil @ 2g/kg seed.
- Drench the base of stem with Copper oxy chloride @ 0.25% or Metalaxyl @ 0.1% or Bordeaux mixture @ 1.0%.

2. POWDERY MILDEW

Oidium caricae

Economic Importance

- Powdery mildew is widely prevalent in Maharashtra and Karnataka.

Symptoms

- White powdery growth on the under surface of the infected leaves.
- Leaves show blotches of yellow or pale green look near the veins at the infection spots which are surrounded by normal green colored tissues.
- Under severe attacks, the top portion of the seedlings may die and the fruit yield is reduced.

Pathogen

- It is an obligate parasite.

- The mycelium is hyaline, septate and haustoria develop in epidermal cells.
- Conidia are hyaline.

Disease Cycle

- The fungus survives as mycelium or cleistothecia in infected plant debris.
- The fungal mycelium growing on the under surface of leaf produces chains of spores that are carried by wind and infected healthy leaves.
- These sproes germinate, send haustoria in to the leaf, develop a fungal vegetative body and reproduce the disease cycle.

Favourable Conditions

- Maximum disease incidence during Sept - Nov with a peak in October.

Management

- Spray wettable Sulphur @ 0.3% at 10 days interval.
- Spray fungicides like Bayleton @ 0.1% or Carbendazim @ 0.1% Karathane 0.05% or Tridemorph @ 0.1% at monthly intervals.

3. ANTHRACNOSE

Colletotrichum gloeosporioides (P.S.–*Glomerella cingulata)*

Economic Importance

- Anthracnose is one of the major disease of papaya in India.
- It affect fruits and leaves in most of the papaya growing areas.
- Anthracnose rot attains serious status during transit and storage causing considerable economic losses.

Symptom

- Spots on fruits appear as brown superficial discolouration of the skin.
- Spots develop into circular, slightly sunken areas and 1 to 3 cm in diameter.

- Lesions coalesce and sparse mycelial growth appears on the margins of the spots.
- Early fruit infection results in mummification and deformation.
- Necrotic spots on leaves and stems.
- Older parts are generally highly susceptible.

Disease Cycle

- The pathogen is able to survive saprophytically for a long duration on fallen leaves, petioles and fruits.
- Primary Infection: Incipient infection carried by fruits from the field.
- Secondary Infection: Conidia by wind and rain splash.

Favourable Conditions

- Excessive moisture.
- Older leaves and mature fruits are more susceptible.

Management

- Removal of infected fruits and leaves, burning them or burying them in the ground.
- Foliar spray twice with Carbendazin @ 0.1% or Chlorothalonil @ 0.2% at 10 - 15 days interval.
- Fumigation of fruits with benzylisothiocyanate and coating with groundnut oil controls post harvest spots and rots.

4. RING SPOT

Papaya ring spot virus

Economic Importance

- Occurs in a severe form in Central India.
- In India, 1st reported from Mumbai and Pune in 1947.
- 90-100% damage occurs in severe cases.

Symptoms

- Vein clearing, puckering and chlorophyll leaf tissues lobbing in.
- Margin and distal parts of leaves roll downward and inwards, mosaic mottling, dark green blisters, leaf distortion which result in shoe string system and stunting of plants.
- On fruits, circular concentric rings are produced. If affected earlier no fruit formation.
- Vectored by aphids *Aphis gossypii, A. craccivora* and also spreads to cucurbits not through seeds.

Pathogen

- The virus particles are rod shaped and thermal inactivation point of the virus lies between 54 and 60°C.

Mode of Spread

- Vectored by aphids *Aphis gossypii, A. craccivora* and also spreads to cucurbits not through seeds.

Management

- Raise papaya seedlings under insect-proof conditions.
- Plant disease free seedlings.
- Raise sorghum / maize as barrier crop before planting papaya.
- Rogue out affected plants immediately on noticing symptoms.
- Donot raise cucurbits around the field.
- Spray with 2% groundnut oil emulsion weekly.

5. LEAF CURL

Papaya leaf curl virus

Economic Importnace

- Leaf curl of papaya is equally important virus disease next to ring spot virus particularly in northern parts and as important leaf crinkle in southern parts of India.

- It was first repoted in 1939 in Tamil Nadu State and later in Bihar, Delhi, Karnataka, Kerala and Andhra Pradesh
- It results in very poor or even no fruit yield by infected papaya plants.

Symptoms

- Curling, crinkling and distortion of leaves, reduction of leaf lamina, rolling of leaf margins inward and downward, thickening of veins.
- Leaves become leathery, brittle and distorted. Plants stunted. Affected plants does not produce flowers and fruits.
- Spread by whitefly *Bemisia tabaci*.

Survival and Spread

- **Primary Infection:** Infected plant parts.
- **Secondary Infection:** Whitefly, *Bamesia tabaci* and grafting.
- Virus also infects tobacco, tomato, sunhemp, chilli, *Petunia, Zinnia, Datura stramonium*, hollyhock, cape gooseberry etc.
- White fly is considered to be the main transmission agent of the virus in nature.

Management

- Disease free seedlings
- Rouging and destruction of diseased plants
- Vector control with Monochrotophos @ 0.05% or Dimethoate @ 0.2% or Metasystox @ 0.1%.

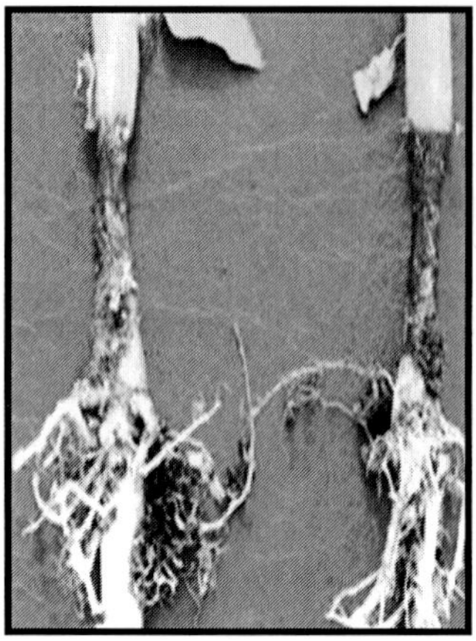

Collar rot

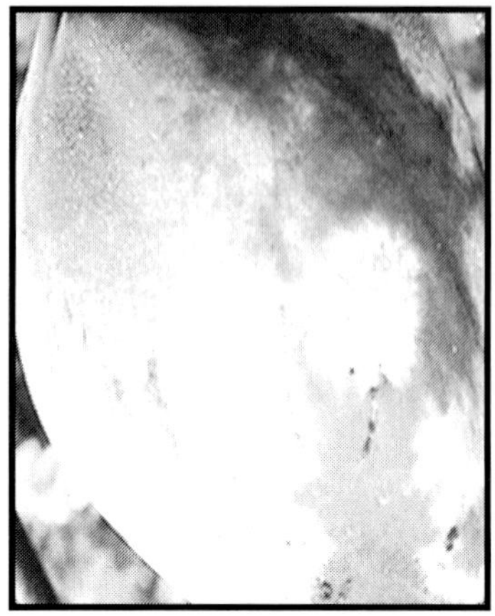

Powdery mildew

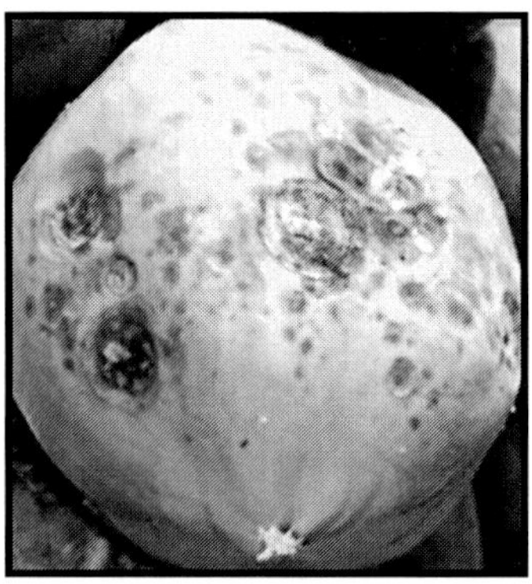

Anthracnose

Ring spot

Leaf curl

CHAPTER - 7

Diseases of Apple and Their Management

SI.No	Major diseases	Causal organism
1	Scab	*Venturia inaequalis*
2	Powdery mildew	*Podosphaera leucotricha*
3	Fire blight	*Erwinia amylovora*
4	Soft rot	*Penicillium expansum*
5	Bitter rot	*Glomerella cingulata*

1. SCAB

Venturia inaequalis (Perfect stage-*Spilocaea pomi)*

Economic Importance

- Scab is considered the most important disease of apples.
- The disease was first reported from Sweden (1819).
- In India, the disease was first reported on Ambri variety in Kashmir valley (1935).
- Scab has now been recorded from all parts of Himachal Pradesh, Uttar Pradesh hills and other areas where apples grown.
- Its primary effect is the reduction of quality of infected fruit, but it also reduces fruit size.
- Losses from apple scab may be 70% or more of the total fruit value.

Symptoms

- Scab infections usually noticed on leaves and fruits.
- Affected leaves become twisted or puckered and have black, circular spots on,their upper surface.
- On the under surface of leaves, the spots are velvety and may coalesce to cover the,whole leaf surface.
- Severely affected leaves may turn yellow and drop.
- Scab can also infect flower stems and cause flowers to drop.
- Scabby spots on fruit begin as sooty, gray-black lesions and may have a white armred halo.
- The lesions later become sunken and tan and may have spores around their margins.
- Infected fruit become distorted and may crack, allowing entry of secondary organisms.
- Severely affected young fruit may drop.

Pathogen

- The mycelium is septate, branched, subcuticular.
- Conidiophores are short, erect, brownish which gives rise to one or two celled, reddish brown conidia.
- Perithecia are dark brown to black with a slight beak and distinct opening (ostiole).
- The perithecia contains usually 50-100 asci .
- Each ascus contains eight bicelled ascospores.

Disease Cycle

- The fungus has two distinct stages in the life cycle ;the saprophytic pseudothecial or over wintering stage on dead fallen leaves, and the conidial, summer stage which is parasitic on leaves, flower buds, fruits, and shoots on the tree during spring and summer.
- The saprophytic stages is considered the major source of survival of the pathogen during dormancy of the host in winter and provides the primary inoculums in the form of ascospores for primary infection of new leaves and flower buds in spring.

- This initiates the active parasitic stage by formation of lesions on leaves in which millions of conidia are produced.
- These conidia are responsible for repeated summer cycles of the life history of the pathogen and spread of the disease during the entire summer.
- Due to short generation time of 8-10 days under ideal conditions of humidity and temperature several cycles (conidia-infection-conidia) occur within 3-4 months and epidemics develop.
- With leaf fall in winter, the pathogen enters its saprophytic stage on dead leaves where ascocarps develop and mature towards the end of the winter just before activity of the tree starts in spring.
- The discharge of ascospores start at a time when the buds on the trees are about to burst open in spring

Environmental Conditions

- Heavy snow and rains in late winter favour the development of ascocarps and the discharge of ascospores.
- Optimum temperature for ascocarp maturity is 16^0C.
- Optimum temperature for vegetative growth of fungus is 19-20^0C
- Continuous sunshine is lethal of ascospores.

Management

- Fallen leaf should be collected and destroyed in the orchard.
- Resistant varieties: Emira, Red free, Ambstraking, Ambroyal, Ambrich and Ambred
- Spray Captan @ 0.2% or Dodine @ 0.25% at short intervals after petal fall.
- Single application of Difolaton @ 0.3% at green bud stage followed by Captan @ 0.2% at petal fall.

Spray Schedule

- 1st spray - Silver tip stage - 0.2% Captofol (or) 0.3% Captan.
- IInd spray-Pink bud - 0.2% Captan (or) 0.3% Mancozeb.

- IIIrd spray - at petal fall - 0.5% Carbendazim.
- IVth spray - after 10 days - 0.2% Captan.
- Vth spray - 15 days after - Mancozeb (0.25%).

Add stickers – teepol or triton 6 ml/10 lit of spray fluid.

2. POWDERY MILDEW

Podosphaera leucotricha.

Economic Importance

- Powdery mildew commonly occurs in the orchards of Apple in Northern India.
- Powdery mildew is a serious disease in nurseries and grown up trees and attacks current year growth of the plant.
- The disease was first reported on apple seedlings in 1877 from USA.
- It considerably damages the leaves and young shoots of the apple if conditions are favorable and , as a result, the fruit yield is reduced.

Symptom

- Small patches of white powdery growth appear on upper side of leaves.
- In severe case the symptom appear on both the sides.
- Twigs are also infected. Affected leaves fall off in severe infection.
- Fruit buds are also affected and deformed or remain small.

Pathogen

- The pathogen is a biotroph.
- Its ectophytic hyphae are hyaline, profusely branched, and develop haustoria in the epidermal cell of host.
- Conidia are ellipsoidal, truncate and hyaline.
- Cleistothecia are subglobose, are densely gregarious, and rarely scattered, and have apical and basal appendages.

- Each cleitothecium contains a single ascus.
- Eight ascospores are present in the ascus.

Disease Cycle

- The mildew fungus over winters mainly as mycelium in dormant blossom and shoot buds produced and infected the previous growing season.
- Conidia are produced and released from the unfolding leaves as they emerge from infected buds at about tight cluster stage.
- Abundant sporulation from over wintering shoots and secondary lesions on young foliage leads to a rapid buildup of inoculum.
- Secondary infection cycles may continue until susceptible tissue is no longer available.
- Since leaves are most susceptible soon after emergence, infection of new leaves may occur as long as shoot growth continues.
- Fruit infection occurs from pink to bloom. Over wintering buds are infected soon after bud initiation.

Favourable Conditions

- Optimum temperature for conidia germination- 19-22°C.
- Relative humidity-60-70%
- Germination does not occur in free moisture.
- Early-season mildew development is affected more by temperature than by relative humidity.

Management

- Sanitation of orchard.
- Pre-bloom spray of lime sulphur (1:60).
- Spray Karathane @ 0.05% or Bitertenol.
- Resistant varieties: Maharaja chunth and Golden Chinese (apple cultivars),Yantarka Altaskya, Dolgoe (Crab apple cultivars)

3. FIRE BLIGHT

Erwinia amylovora

Economic Importance

- Fire blight of apple and pears is one of the most serious diseases of orchard plants.
- It was noticed in the Hudson valley as early as in 1780.
- In 1878, Burrill in the USA established the bacteriological nature of the disease-one of the early discoveries of bacteria as plant pathogens.
- The disease is not yet very serious in Inida

Symptom

- The initial symptom usually occurs on leaves, which become water soaked, then shrivel turn brownish to black in colour and fall or remain hanging in tree.
- The symptom spread to twigs. Terminal twigs wilt from tip to downward and also spread to branches.
- Fruits becomes water soaked, turns brown, shrivels and finally becomes black.
- Oozing may be seen in the affected area.

Pathogen

- The bacterium is rod shaped and motile by peritrichous flagella.
- Bacterium occurs usually singly but pairs or chain of 3 or 4 bacteria also exist.
- Each bacterial cell is enclosed in a capsule.

Disease Cycle

- The bacteria overwinter at the margins of the cankers and possibly in buds and apparently healthy woody tissue.
- In spring, the bacteria in these holdover cankers become active again, multiply and spread into the adjoining healthy bark.

- Bacterial ooze on cankers is carried into open blossoms by rainfall and more commonly by bees, flies and ants.
- Multiplication of the bacteria in the floral nectarines provides secondary inoculums which may spread by rain splash and insects.

Favourable Conditions

- The spread of the pathogen is favored by warm, humid weather.
- Optimum temperature for disease development-18.5^{0}C.

Management

- During winter prune out and burn blighted twigs, branches and cankers, at about 30cm below the point of visible infection.
- Disinfect the tools after each cut with 10% sodium hypochlorite.
- Resistant varieties.
- Dormant sprays with Copper sulphate or Bordeaux mixture, however, Bordeaux mixture or streptomycin are the only effective blossom sprays.

4. SOFT ROT

Penicillium expansum

Economic Importance

- Soft rot is a storage and transit disease of apple fruits and is found throughout the world.
- It can cause 75 percent loss of fruits during transit and storage.
- The disease is also called Penicillium rot, Bin rot and Blue mould.

Symptom

- The rottened area of the fruit looks light or yellowish brown, soft and watery.
- Skin becomes wrinkled.

- A peculiar musty odour is emitted.
- Under humid condition a bluish green sporulating growth appears.
- Infection take place by wounds in the skin caused by insects and during handing in storage and transport.

Pathogen

- The aerial mycelium is colorless, although the growth after sporulation look blue or bluish green.
- The conidiophores arise in clusters, forming definite fascicles. Conidiophores give rise to 1-3 main branches.
- They in turn produce crowded whorls of branches.
- Conidia are formed in chains.

Disease Cycle

- Spores of the soft rot fungus are present almost everywhere and can survive long periods of unfavorable conditions
- The fungal spores are spread by air.
- Mycelium can infect through bruised or wounded fruits in close contact.
- Infection of the fruit usually take place through wounds in the skin, such as wounds caused by insect bites, careless picking, rough handling during washing, grading , packing, transit and storage.
- Infection is also through lenticels.
- At ordinary temperatures, infected fruit can rot in 2 weeks or less storage and transport.

Management

- Careful handling of fruits without causing any wounds.
- Dipping the fruits in aureofunginsol @ 500 ppm for 20 min gives best control.

5. BITTER ROT

Glomerella cingulata

Symptom

- Faint, light brown discolouration beneath the skin develops. The discolouration expands in a cone shape. The circular, rough lesions become depressed. The lesions increased and covers entire areas of fruits.
- Diny black dots appear beneath the cuticle which gives rise to acervuli.
- Pink masses of spores are found arranged in defined rings.

Disease Cycle

- The fungus over winters in mummified fruit, in cracks and crevices in bark, and in cankers produced by the bitter rot fungus or by other diseases, such as fire blight. Jagged edges of broken limbs are also ideal sites.
- The bitter rot fungus is one of the few rot organisms that can penetrate unbroken skin of fruit.
- Although penetration is direct, wounds can be colonized rapidly by the fungus.
- Spores are waterborne and are released during rainfall throughout the growing season. Fruit infection can occur early but is more common from mid to late season.
- Often, the first infections appear in cone-shaped areas within the tree beneath mummies or a canker.

Environmental Conditions

- Optimum temperature-26^0C
- Relative humidity - 80 to 100 percent,

Management

- Spray Mancozeb (0.25%) in field.
- Treatment with Mancozeb (0.25%) to check the disease.

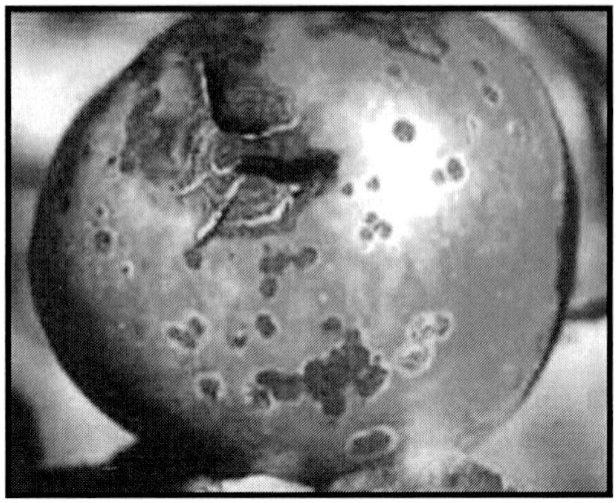

Scab

Powdery mildew

Fire blight

Chapter - 8

Diseases of Pomegranate and Their Management

SI.No	Major diseases	Causal organism
1	Leaf Spot or Blight	*Colletotrichum gloesporioides*
2	Cercospora leaf & fruit Spot	*Cercospora* sp
3	Alternaria fruit spot	*Alternaria alternata*
4	Bacterial leaf spot	*Xanthomonas axonopodis pv. punicae*

1. LEAF SPOT OR BLIGHT OR ANTHRACNOSE

Colletotrichum gloesporioides

Symptom

- Characterized by appearance of small, irregular and water-soaked spots on leaves.
- Affected leaves fall off.
- Small irregular lesions on the fruit.
- Reduction in market value.

Pathogen

- Conidiophores are brown, short, fasciculate, sparingly septate.
- Conidia are hyaline to pale olivaceous, cylindric and septate.

Disease Cycle

- The pathogen spread through wind borne conidia.

Favourable Conditions

- High humidity and temperature ranging from 20 to 27°C.
- Severe during August-September.

Management

- Spray Carbendazim (0.1%) or Mancozeb (0.2%) at 15 days intervals.

2. CERCOSPORA LEAF AND FRUIT SPOT

Cercospora sp.

Symptoms

- Light zonate brown spots appear on the leaves and fruits.
- The affected areas in the twigs become flattened and depressed with raised edge.
- The affected fruits showed small irregular black spots, which later on coalesce, into big spots.
- In severe cases the whole plant dies.

Disease Cycle

- Primary Infection: Diseased plant debris
- Secondary Infection: Wind borne conidia

Management

- The diseased fruits should be collected and destroyed.
- Two to three spray with Mancozeb (0.25%) at 15 days interval.

3. ALTERNARIA FRUIT SPOT

Alternaria alternata

Symptom

- Small reddish brown circular spots appear on the fruits.
- Under favorable conditions, these spots, coalesce to form larger patches and the fruits start rotting.
- The arils get affected which become pale and become unfit for consumption.

Management

- All the affected fruits should be collected and destroyed.
- Spraying Mancozeb (0.25 %) effectively controls the disease.

4. BACTERIAL LEAF SPOT

Xanthomonas axonopodis pv. punicae

Symptoms

- Small irregular water soaked spots,2-5 mm in diameter appear on the leaves.
- Leaves distorted and malformed.
- Severely infected young leaves shed.
- The bacterium attacks stem, branches and fruits also.
- On the stem, the disease starts as brown to black spots around the nodes and finally girdling and cracking of nodes.
- Branches break down.
- Brown to black raised spots on the fruit with L or Y shaped cracks.

Disease Cycle

- Primary Infection : Infected cuttings.
- Secondary Infection: Wind splashed rains.
- The bacteria infect through wounds and stomatal openings.

Management

- Clean cultivation and strict sanitation in orchard.
- Spray Carbendazim (1%) along with antibiotic.

Cercospora fruit spot

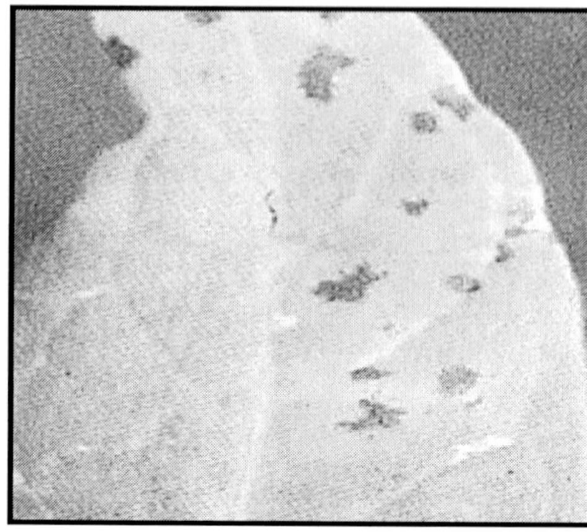

Leaf Spot or Blight

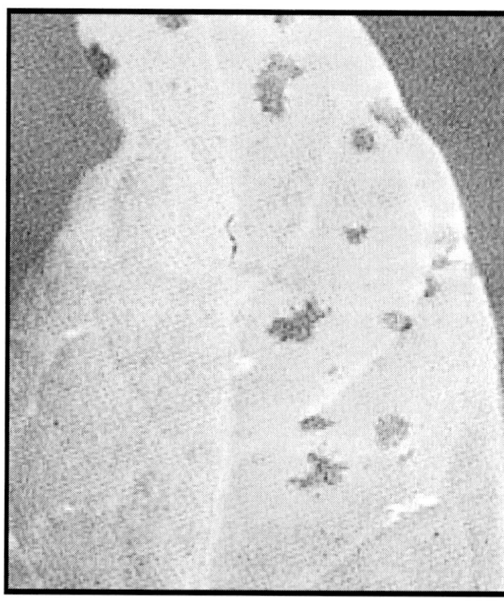

Bacterial leaf spot

CHAPTER - 9

Diseases of Sapota and Their Management

SI.No	Major diseases	Causal organism
1	Sooty mould	*Capnodium* sp.
2	Leaf spot	*Phavophleospora indica*
3	Flat limb	*Botryodiplodia theobromae*

1. SOOTY MOULD

Capnodium sp.

Symptom

- Shooty mould is a fungal disease developed on honeydew-like excretion secreted by aphids and scale insects.
- The fungus slowly covers the entire leaf area severely affecting the process of photosynthesis.
- This results in reduced translocation of food to the fruits, which leads to reduction in their size.

Management

- Spray starch solution (5%) to remove the fungal growth.
- Control insects by spraying systemic insecticides.

2. LEAF SPOT

Phaeoleospora indica

Symptom

- Characterized by numerous, small, pinkish to reddish brown spots with whitish centers in the leaf.
- Spots coalesce and leaves drop prematurely.

Favourable Conditions

- Severe during October-December.
- Fungus grows best at 25^0 C temperature and 90% relative humidity.

Management

- Spray Mancozeb (0.25%) or Copper oxychloride (0.3%) at 15 days intervals.

3. FLAT LIMB

Botryodiplodia theobromae

Symptoms

- Leaves become thin, small and yellow.
- Branches of affected trees become flat and twisted.
- Cluster of leaves and flowers on affected twigs.
- If fruits are set, they are undersized, hard and fail to ripen.
- Foliage and fruits fall prematurely.

Management

- Pruning of affected branches followed by Mancozeb (0.25%) spray.

Sooty mould

Leaf spot

Flat Limb

Diseases of Plantation Crops

Chapter - 10

Diseases of Coconut and Their Management

SI.No	Major diseases	Causal organism
1.	Bud rot	*Phytophthora palmivora*
2.	Stem bleeding	*Ceratocystis paradoxa/ Ophiostoma paradoxa*
3.	Grey blight	*Pestalotiopsis palmarum*
4.	Basal Stem Rot (Thanjavur wilt/ Bole rot)	*Ganoderma lucidum*

1. BUD ROT

Phytophthora palmivora

Economic Importance

- First reported by Butler in 1906.
- Coconut and toddy palm (*Borassus flabellifer*) suffer badly.
- Also infects Palmyrah & Arecanut.
- Datepalm is immune to the disease.
- Bud rot affects the palms at various stages of growth.

Symptoms

- Severe on young palms.
- Yellowish green discoloration of the heart leaf or crown leaf.

- The basal tissues of the leaf rot quickly and can be easily separated from the crown.
- Spindle withers and droop down.
- Older leaves develop irregular water soaked spots which are sunken in nature.
- The leaves and sheath in the central spindle fall off leaving an outer whorl of green leaves.
- The withered central shoot can be pulled out very easily from the crown.
- The central crown may rot and in few months the tree may wilt.
- Young nuts fail to mature and fall.

Pathogen

- The fungus produces intercellur, non septate, hyaline mycelium.
- Sporangiophores are hyaline and simple or branched occasionally.
- The sporangiophores are hyaline, thin walled, pear shaped with a prominent papillae.
- Sporangia releases reniform, biflagellate zoospores upon germination.
- The fungus also produces thick walled, spherical oospores.
- In addition, thick walled, yellowish brown chlamydospores are also produced.

Disease Cycle

- Primary Infection: Through dormant, mycelium oospores or chlamydospores carried over summer months on the host debris. With the onset of monsoon rains, the fungus becomes active producing cottony mycelium that infects tender host tissue.
- Secondary Infection: Through sporangia with numerous zoospores which spread rapidly in the rain water or sporangia spread through wind and insects also.

Favourable Conditions

- High rainfall, high atmospheric humidity (above 90 per cent), low temperature (18-20°C) and wounds caused by tappper and rhinoceros beetles.

Management

- Cutting and burning of badly infected palms.
- If the disease is detected early, remove the infected portions and protect with Bordeaux paste (Tree surgery).
- Spray copper fungicides (Copper oxy chloride) after on set of monsoon to prevent infection.
- Give prophylactic spray with 1% Bordeaux mixture to all the healthy palms in the vicinity of disease one and also before onset of monsoon rains.

2. STEM BLEEDING

Ceratocystis paradoxa/Ophiostoma paradoxa

Symptoms

- The characteristic symptom is the exudation of reddish brown fluid from the cracks in the stem.
- The fluid trickles down to several feet on the stem and the exudates dries up forming a black crust. The tissues below the cracks turn yellow and decay. As the disease progresses, more area underneath the bark gets decayed and the bleeding patch extends further up.
- The vigour of the tree is affected and nut yield is reduced.
- The tree is not killed out right but become uneconomical to maintain.
- In extreme cases, the trees may become barren and die.

Pathogen

- The fungus produces two types of conidia.
- Macroconidia are produced on conidiophores, either singly or in chains.

- They are spherical or dark green in colour.
- Microconidia are produced endogeneously inside the long cells of conidiophores and cell ruptures when mature and release the microconidia in long chain.
- The pathogen also produces hyaline perithecia with clavate asci and hyaline ascospores.

Disease Cycle

- Primary Infection: Through perithecia and chlamydospores in cracks and cavities of infected stem.
- Secondary Infection: Through conidia dispersed by irrigation water, insects, wind.

Favourable Conditions

- Heavy rainfall followed by drought.
- Cracks and insect wounds on stem.
- Physiological imbalances.
- Poor fertilization and poor maintenance of orchard.
- Damage by *Diocalandra* and *Xyleborus* beetles.

Management

- Avoid any mechanical injury to the stem.
- Provide proper drainage during rainy season.
- Chisel out affected tissues and dress the wound with hot coaltar or Bordeaux paste.
- In older trees, after draining the fluid, cavities can be stuffed with a mixture of coaltar and sawdust.
- Cover the lower part of trunk with coal tar or Bordeaux paste to prevent the disease.
- Along with 50kg of organic manure, apply 5 kg neem cake containing the antagonistic fungi, *Trichoderma* to the basin during September.
- Coconut stem boring insects should be controlled by applying Carbaryl 50% WP on the trunk at 3g per liter.

3. GREY BLIGHT

Pestalotiopsis palmarum

Symptoms

- Disease symptoms develop in mature leaves of outer whorl.
- Minute yellow spots surrounded by a grayish margin appear on leaflets. Gradually, the centre of the spots turns grayish white with dark brown margins and a yellow halo.
- Spots coalesce into irregular necrotic patches.
- Complete drying and shrivelling of leaf blade occur giving a blighted or burnt appearance.
- Large number of globose or ovoid black acervuli appears on the upper surface of affected leaves.

Pathogen

- Mycelium is septate, branched, light brown, inter and intra cellular.
- Fungus produces acervuli as its asexual fruiting body during sporulation.
- Acervuli are black, cushion shaped, sub epidermal and break open to expose conidia and black sterile structures called setae.
- Conidiophores are hyaline, short, simple and bear a conidium at the tip.
- Conidia are five celled, middle three cells are coloured, basal and tip cells are hyaline. Tip cells have 3 -5 slender elongated appendages.

Disease Cycle

- Primary Infection: Through conidia in plant debris.
- Secondary Infection: Through conidia carried by wind splashed rain water.

Favourable Conditions

- Well drained soils, soils with potash deficiency, continuous rainy weather for 4-5 days and strong winds.

Management

- Remove older affected leaves and burn.
- Apply heavy dose of potash and improve drainage condition of soil.
- Spray foliage with Copper oxychloride @ 0.3% or Captan @ 0.2%.

4. BASAL STEM ROT (THANJAVUR WILT / BOLE ROT)

Ganoderma lucidum

Symptoms

- The most usual symptoms are yellowing, withering and drooping of the outer fronds which remain hanging around the trunk for several months before shedding.
- The younger leaves remain green for sometime and later turn yellowish brown.
- The new fronds produced become successively smaller and yellowish in colour which do not unfold properly.
- The wilting plants also show bleeding patches near the base of the trunk.
- A brown gummy liquid oozes out from the cracks in the tree which slowly result in the death of outer tissues.
- As the infection advances, fresh bleeding patches appear above the old once, up to 3-5 meters height. The decay of the basal portion occurs slowly and tree succumbs to the diseases in 2-3 years.

Pathogen

- The fungus produces a semi circular basidiocarp (bracket), which is attached to the tree with a stalk.
- The bracket is very big about 10-12 cm diameter and woody.
- The upper surface is tough, shining, light to dark brown or almost black with concentric furrows.
- The lower surface is white and soft with numerous minute pores.

- These pores represent the opening of the hymenial tubes, which are lined with basidia and basidio-spores.
- Basidiospores are oval, brown and thick walled.

Favourable Conditions

- Trees grown in sandy loam and sandy soils, water logging during severe rains, low soil moisture content during summer months and damages caused by weevils and beetles.

Disease Cycle

- The fungus is soil-borne and survives in the soil for long time.
- The primary infection is through basidiospores in the soil, which attack roots.
- The irrigation water and rain water also help in the spread of the fungus.

Management

- Remove and burn severely infected trees which are beyond recovery.
- Isolate the diseased trees by digging a trench all around to check further spread. Irrigate the palms at least once in a fortnight during summer months.
- Apply heavy doses of farm yard manure or compost or green manure at 50 Kg/tree/year along with 5 kg of neem cake.
- Drench the soil near the tree with 40 litres of 1 per cent Bordeaux mixture at quarterly interval for thrice a year and repeat after 2-3 years.
- Apply Aureofunginsol 2g + Copper sulphate 1g in100 ml of water or Tridemorph 2ml/100 ml of water through stem injection or root feeding at quarterly intervals.

Bud rot

Basal stem rot

CHAPTER - 11

Diseases of Coffee and Their Management

Sl.No	Major diseases	Causal organism
1.	Coffee rust/Oriental leaf disease	*Hemileia vastatrix*
2.	Black rot or Koleroga	*Cortieium salminicolor*
3.	Die back or Anthracnose	*Colletotrichum coffeanum*
4.	Damping off / Collar rot	*Rhizoctonia solani*
5.	Berry blotch	*Cercospora coffeicola*

1. COFFEE RUST/ORIENTAL LEAF DISEASE

Hemileia vastatrix

Economic Importance

- Rust of coffee is not only one of the most dangerous diseases of coffee but ranks as one of the classical plant diseases of international significance.
- It was recorded in Sri Lanka in 1968 and by 1993 the disease assumed such severe form that the coffee plantations in Sri Lanka were replaced by tea cultivation.
- In India, the disease was first recorded in the year 1870 in Karnataka and occurs every year, threatening coffee cultivation.
- It occurs in Karnataka, Kerala and even in Madhya Pradesh where coffee is grown in a small tract.

- The worst damage is usually caused during the rains.

Symptoms

- The disease is restricted to leaves although sometimes it can seen on tender shoots and berries.
- The rust pustules appear as small spots, 1-2 mm in diameter.
- In early stages these spots are yellowish becoming orange in color with increase in size.
- On the upper surface, opposite the spots on the lower surface, the color is often brownish.
- In severe attack leaves may dry and wither.
- The berries remain small and fail to ripen.
- Extensive defoliation weakens the trees and results in poor yield, severe die back of twigs, and death of trees.

Pathogen

- The mycelium is intercellular and sends haustoria into the cells.
- The mycelium sends out erumpent stalks through stomata which bear the uredospores.
- The uredospores are reniform or orange segment like in shape.
- The convex side of the spores are echinulated and the lower side is smooth and measure 26 – 40 x 20 – 30 micron meter.
- The telial stage succeeds the uredial stage in the later stage.
- The pycnial and aecial stages of *H. vastatrix* has not been seen and it is unknown whether the fungus is autoecious or heterocious

Disease Cycle

- The fungus survives primarily as mycelium, uredia and uredospores on infected leaves.
- The uredospores are easily disseminated by wind, rain and possibly insects.
- Under favourable conditions the spores germinate and enter the host through stomata.

- The mycelium grow intercellularly sending haustoria into the cells.
- Repeated crops of uredospores are produced during the growing season and , as a result , several secondary infection cycles take place if environmental conditions are favourable.
- Young leaves are more susceptible than older leaves.

Favourable conditions

- Rainy weather, heavy dew and high temperature favours disease incidence.
- Uredospores germinate only in free moisture.

Management

- Fallen leaves should be composted or destroyed.
- Spray the bushes once with Bordeaux mixture @ 0.5% or Copper oxy chloride @ 0.25% and subsequently 2-3 times during monsoon.
- Spray Triadimefon @ 0.05%.

2. BLACK ROT OR KOLEROGA

Cortieium salminicolor

Economic Importance

- In India, it occurs in Karnataka and Tamil Nadu.
- In South India the disease is severe only in those areas growing with *C. arabica.*
- It is influenced by south west monsoon period from June - September.

Symptoms

- Dark brown or black decaying of leaves, twigs and berries.
- Affected leaves get detached and hang down by means of slimy fungal strands.
- Blackening and rotting of affected leaves, young twigs and berries.
- Defoliation and berry drop occur.

Pathogen

- The hyphae are hyaline when young and turn light brown with age.
- Fructifications arise with numerous basidia and basidiospores.
- Basidia are simple, oval rounded or pyriform.
- Basidiospores are hyaline, elongated, rounded at one end, slightly concave on other side.
- Fungus forms sclerotia or hyphal clumps at later stage by repeated branching of short cells.

Disease Cycle

- The pathogen spread through infected plant debris.
- Mycelium lies in twigs throughout year.
- The pathogen penetrates the leaves through the stomata on the lower side and the hyphae invade intercellularly in the palisade tissue.
- The fungus mostly spreads by contact from leaf to leaf through the vegetative mycelium.

Management

- Remove and burn affected parts.
- Apply 1% Bordeaux mixture close to the south westerly monsoon if needed.
- Centre the coffee bushes, regulate the overhead canopy.

3. DIE BACK OR ANTHRACNOSE

Colletotrichum coffeanum

Symptoms

- Circular to grayish spots of 2-3 mm in diameter appear on leaves.
- Small dark coloured sunken spots are formed on berries.
- Beans become brown.
- Die back also occurs.

Disease Cycle

- The pathogen occurs as a saprophyte on dead tissue on the outer layer of the bark, which provides the major source of inoculum.
- It release large numbers of water borne conidia during the wet season.
- Conidia are spread by rain water percolating through the canopy and rain splash can disperse conidia between trees.
- Long distance dispersal occurs primarily by the carriage of conidia on passive vectors such as birds, machinery etc.

Management

- Provide good drainage to remove excess water from the root zone.
- Spray with 0.5% Bordeaux mixture or Mancozeb (0.25%) to prevent defoliation.
- Maintain adequate shade to avoid sunscald.

4. DAMPING OFF / COLLAR ROT

Rhizoctonia solani

Symptoms

- It caused pre emergence damping off and post emergence damping off.
- In post emergence damping off, collar region near soil level is infected leading the rotting of tissue and death of seedlings.

Disease Cycle

- The disease is soil borne.

Management

- Soil drenching with Copper oxychloride (0.25%).
- Maintain good shade and sanitation.
- Remove and burn affected seedlings.

5. BERRY BLOTCH

Cercospora coffeicola

Symptoms

- Necrotic spots on the exposed surface of green berries enlarge and cover the major portion.
- Fruit skin shrivels and sticks fast.

Pathogen

- Conidiophores are short, fasiculate and olivaceous.
- Conidia are subcylindrical, hyaline, 2-3 septate and 40-60x 3.5 micron meter in size.

Disease cycle

- The pathogen is seed borne.
- Conidia are spread by wind.

Management

- Spray 1% Bordeaux mixture during june and late august.
- Maintain medium shade overhead for one year.

Rust

Brown leaf spot

Anthracnose

CHAPTER - 12

Diseases of Tea and Their Management

SI.No	Major diseases	Causal organism
1.	Blister blight of tea	*Exobasidium vexans*
2.	Red rust	*Cephaleurus mycoidea*
3.	Black rot	*Corticium invisum* and *C. theae*
4.	Black root	*Rosellina areuata*

1. BLISTER BLIGHT OF TEA

Exobasidium vexans

Symptoms

- Initially oily, yellowish, translucent spots appear on the tender leaf and turn to deep red shiny blisters.
- The circular spot gradually enlarges to 3 to 13 mm diameter, bulged on the under surface of the leaf with a concave trough like depression on the upper surface.
- Leaves become curled and distorted.
- First flush of 2-3 young leaves are attacked and the young shoots and buds are killed.
- Mature leaf is not affected.
- In nursery infection, seedlings are stunted with many thin stems instead of a single stalk.

- Repeated attacks cause death of seedlings.

Pathogen

- The mycelium is confined to the blistered areas on the leaves.
- They are septae and collect in bundles below the lower epidermis.
- Later by rupturing the epidermis a continuous layer of vertical hyphae are projected on the surface of spot.
- The fungus produces two kind of spores viz.,the conidia and basidiospores.
- The conidia are most abundant, borne singly at the tip of long stalks.
- Basidia are formed on the surface in large number but never form a continuous hymenium.

Disease cycle

- The fungus completes its life cycle in 11-28 days and several generations of spores are produced in a season.
- It produces conidia and basidiospores in the same blister.
- Spores are air borne.
- The basidiospores germinate on tea leaves by forming appresorium from which infection takes place through an infection hypha.
- It is a polycyclic diseases and the secondary infections are repeated till plants are available in the field.
- The perpetuation of the fungus appears to be from the pre-existing infected bushes.

Favourable Conditions

- Relative humidity more than 83% for 7 to 10 days favours disease.
- Temperature above 35°C inhibit the disease.
- Bushes in low, moist and shady localities suffer more.
- Pruned trees with new flush is highly susceptible.

Management

- Seedlings should be protected in nursery by weekly sprays of 0.3% Copper oxychloride.
- Spray, a mixture of 210g Copper Oxy Chloride + 210g Nickel chloride per ha at 5 days intervals from June-September and 11 day intervals in October-November.
- Mancozeb, Tridemorph, Triadimefon and Pyracarbolid (Sicarol) offers good disease control under field conditions.

2. RED RUST

Cephaleurus mycoidea

Symptoms

- Orange yellow, circular patches appear on upper surface of leaves. The spots become brown and dry up. When it affects the given stem it hardens prematurely.

Pathogen

- *Cephaleurus mycoidea.*

Favourable Conditions

- Rainy season is best suited for propagation of algae.

Management

- Removal of infected portion and spraying with Copper oxychloride (0.25%).

3. BLACK ROT

Corticium invisum and *C. theae*

Symptoms

- Small dark brown irregular spots appear on leaf.
- They coalesce to produce a dark brown patch which eventually covers the whole leaf and drop off.
- Before the leaf turns black the lower surface assumes a white powdery appearance.

Pathogen

- *Corticium invisum* and *C. theae*

Disease cycle

- Basidiospores carried by workers.
- The disease develops rapidly when temperature is high and air is humid.
- At the beginning of rainfall they germinate and produce hyphae which start fresh infection.

Favourable Conditions

- Occur in nursery shaded with Crotalaria.
- Basidiospores germinate only in wet weather or when leaves are covered with dew.

Management

- Prune in December end, remove the prunings immediately, burn after drying.
- Collect all dead and dried leaves.
- Spray a copper fungicide in third week of April.

4. BLACK ROOT

Rosellina areuata

Symptoms

- The fungus originate from the dead heaped leaves of 5 - 7.5 cm above the soil level.
- From there it spreads to roots region of tea bushes.
- Star like growth of mycelium can be seen when bark is removed.
- The mycelium surrounds the stem at the surface of the soil and kills the bark for the length of 7.5 - 10.0 cm.
- A swollen ring of tissue is formed round the stem above the dead patch.

Pathogen

- The pathogen produces two kinds of fructification, a perithecial stage and a conidial stage.
- The conidia are borne on short bristle like stalks.
- The perithecia are black and spherical.
- They bear asci which in turn bear ascospores.

Mode of Spread

- The disease is spread by wind.

Management

- Clean cultivation with out fallen leaves.
- Dig a drench around the infected bush to provide sunlight in the drench which prevent the spread of mycelium.
- Removal of infected portion and spraying with Copper oxychloride (0.25%).

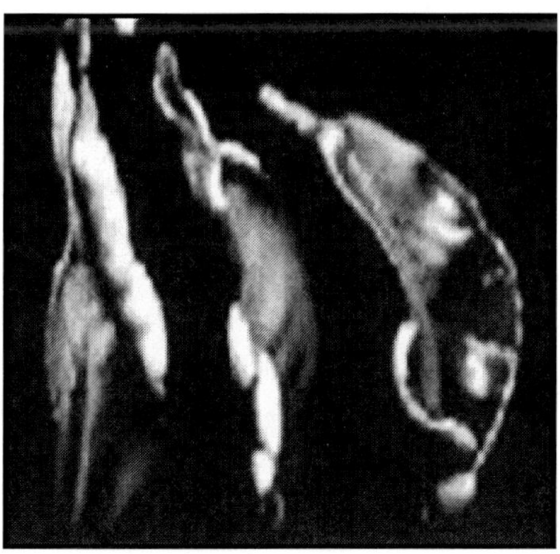

Blister blight

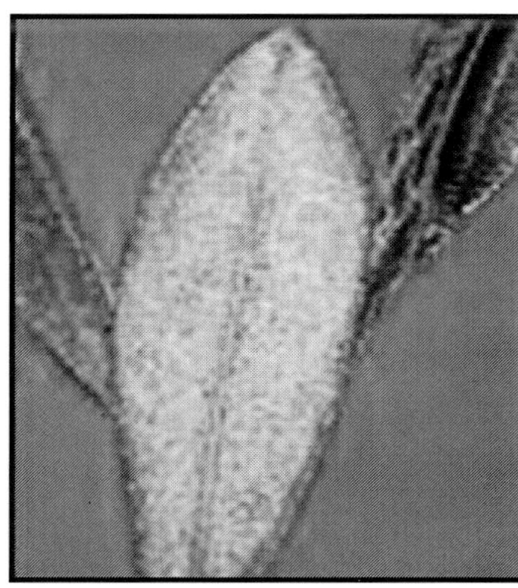

Red rust

CHAPTER - 13

Diseases of Betelvine and Their Management

SI.No	Major diseases	Causal organism
1.	Foot rot or Leaf rot or wilt	*Phytophthora parasitica* var. *piperina*
2.	Anthracnose	*Colletotrichum piperis*
3.	Bacterial leaf spot or stem rot	*Xanthomonas campestris* p.v. *betlicola*

1. FOOT ROT OR LEAF ROT OR WILT

Phytophthora parasitica var. *piperina*

Economic Importance

- This disease is most common in betelvine gardens of India.
- During some years it occurs in epiphytotic form, causing devastating damage to the crops.
- In India during, 1923-1924, it completely wiped out the crop in the Central provinces.

Symptoms

- It is the most destructive fungal disease that produces both a wet rot symptom on leaves and wilt due to foot rot.
- At first circular, dark brown spot appear which becomes wet and rot under continuous high humidity conditions, Otherwise dark brown necrotic spots with alternate light brown zonations appear.

- The loss of lusture of the leaves in the foot rot conditions followed by complete wilting and drying of the vines occur in a quick sequence.

Pathogen

- The fungus produces hyaline, non septate mycelium.
- The sporangia are thin walled, hyaline ovate or lemon shaped with papillae, measuring 30-40 X 15-20um.
- Zoospores, which are liberated from the sporangia, are kidney-shaped and biflagellate. Oospores are dark brown, globose and thick walled.

Disease Cycle

- The fungus is soil-borne and survives as facultative saprophyte in the infected plant debris and in the soil as oospores and chlamydospores.
- The fungus mainly spreads from field to field through irrigation water.
- The secondary spread is through sporangia and zoospores disseminated by splash irrigation and wind-borne rains.

Favourable Conditions

- September to February months with high atmospheric humidity and low night temperature (23°C and below) are highly favourable.

Management

- Select well matured (more than 1 year old) seed vines free from pest and diseases.
- Soak the seed vines for about 30 minutes in 1% Bordeaux mixture.
- Drench Bordeaux mixture (1%) in basins formed around the vine at monthly intervals starting from October – January, three times soil drench and six times spray from June - July.
- During winter season, avoid frequent irrigation.

- Remove the affected vines away from the garden and burn them.
- Application of *Trichoderma viride* @ 5 g/vine.

2. ANTHRACNOSE

Colletotrichum piperis

Symptoms

- The leaf spot symptom is irregular in shape and size, light to dark brown surrounded by diffuse chlorotic yellow halo.
- Marginal leaf tissue becomes black, necrotic and gradually spreads towards the leaf centre. Occasionally diffused yellow halo also develops.
- In the anthracnose stage circular, black lesions that occur rapidly increases in size and girdle the stem culminating in the death of the vine.

Disease Cycle

- The fungus remain in the infected plant debris in the field.
- The primary infection is through the soil-borne conidia, spread by rainwater splash or splash irrigation.
- The secondary spread in the field is aided by air-borne conidia.

Pathogen

- The fungus produces large number of acervuli containing short, hyaline conidiophores and black coloured setae.
- The conidia are single celled, hyaline and falcate.

Management

- Collect and destory the infected vines and leaves.
- Spray 0.2 per cent Ziram or 0.5 per cent Bordeaux mixture after plucking the leaves.

3. BACTERIAL LEAF SPOT OR STEM ROT

Xanthomonas campestris pv *betlicola*

Symptoms

- The bacterium produces circular or elliptical spot with halo on both side of leaves.
- The affected leaves turn yellow and fall of in advance stage.In stem canker, balack spot appears just below the node and stem breaks from the affected parts.
- The infestation spreads in the tissues and thus the stem shrinks and burst resulting in to drying of the upper vine which looks like wilting symptoms.

Pathogen

- Bacterium is a small rod with a single polar flagellum.
- It is Gram negative and non-spore forming.

Disease Cycle

- The bacteria which are viable in the infected vines and leaves serve as a primary source of inoculum.
- Rain splashes and splash irrigation water help in the secondary spread.

Favourable Conditions

- Cloudy weather with intermittent rains and high relative humidity.
- Two to 3 years old vines are highly susceptible.

Management

- Remove all debris and plant parts.
- Cuttings should be dipped in the solution of Streptocyclin sulphate or Agrimycin (0.025%).
- Spray Streptocycline @ 400 ppm + Bordeaux mixture @ 0.25% at the time of first disease symptoms appearance. Continue spraying at 20 days intervals. Always spray the chemical after plucking the leaves.

Leaf spot

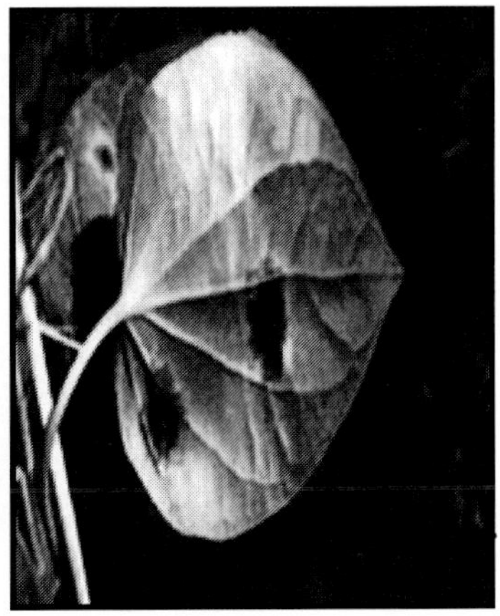

Anthracnose

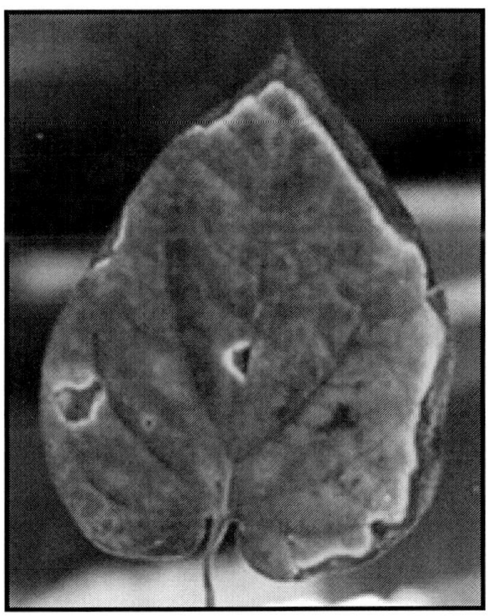

Bacterial blight

Diseases of Vegetable Crops

Chapter - 14

Diseases of Potato and Their Management

SI.No	Major diseases	Causal organism
1.	Late blight	*Phytophthora infestans*
2.	Early Blight	*Alternaria solani*
3.	Black scurf	*Rhizoctonia solani*
4.	Common scab	*Streptomyces scabies*
5.	Brown rot	*Ralstonia solanacearum*
6.	Wart	*Synchytrium endobioticum*
7.	Black leg (Soft rot)	*Erwinia caratovora subsp. caratovora*
8.	Viral diseases	
a.	Mild mosaic/Interveinal mosaic	*Potato virus* X
b.	Severe mosaic	*Potato virus* Y
c.	Leaf roll	*Potato leaf roll virus*
9.	Potato spindle tuber	Viroid

1. LATE BLIGHT

Phytophthora infestans

Economic Importance

- Late blight is the most destructive disease of potato.
- First reported from Andes mountains of South America.

- The famous Irish famine of 1845-46 was due to this disease.
- In India, the disease was first reported in Darjeeling district in India (1880).
- Under favorable conditions, the disease is capable of causing cent percent losses.
- Several epidemics have been reported in India causing loss up to 75 percent.

Symptoms

- Initially starts from leaf tips or margins and spread inward.
- Small faded green patches on upper surface of leaf which turn into brown spots.
- Downy growth of the pathogen on subsequent lower surface.
- Progressive defoliation and collapse of plants under favourable conditions.
- Water soaked stripes on stem which becomes necrotic.
- Purplish brown spots appear on skin of tubers.
- On cutting, the affected tubers show rusty brown necrosis spreading from surface to the centre.

Pathogen

- The mycelium is endophytic, coenocytic and hyaline which are inter cellular with double club shaped haustoria type.
- Sporangiophores are hyaline, branched intermediate and thick walled.
- Sporangia are thin walled, hyaline, oval or pear shaped with a definite papilla at the apex.
- The sporangium may act as a conidium and germinate directly to form a germ tube.
- Zoospores are biflagellate possess fine hairs while the other does not.
- The sexual reproduction in *Phytophthora* is not very common.

Disease Cycle

- Primary infection is through use of infected tubers.

- Mycelium spreads into shoots produced from infected tubers and reaches the aerial parts of the plant.
- Sporangiophore emerges through stomata on stem and leaves and produce sporangia, which are spread by rain to wet potato leaves or stems and cause disease.
- Large number of asexual generation in a growing season kills the foliage rapidly.
- The zoospores found in the soil germinate, penetrate through lentils or wounds into the tubers and send intercellular mycelium and haustoria into the cells and cause infection.

Favourable Conditions

- High relative humidity (>90%) coupled with suitable temperature are the principal factor governing occurrence of late blight epidemics.
- The optimum temperature for germination of the sporangia by zoospores is 12-13^{0}C and by germ tube 24^{0}C.
- Sporulation can occurs at any temperature from 9-26^{0}C.
- The optimum temperature for growth of mycelium is 16-18^{0}C.

Dutch Rules

- Night temperature below the dew point for 4 hours or more.
- Night temperature not below 10°C.
- Cloudiness on the next day.
- Rainfall at least 0.1mm on the following day.

Management

- Regulatory measures.
- Select healthy tubers for planting.
- Delayed harvesting.
- High ridging to about 10-15cm height reduces tuber infection.
- Apply potasic fertilizer.
- Grow resistant varieties such as Kufri Jyothi, Kufri Badshah, Kufri Jeevan, Kufri Sherpa, *etc*.

- Dip sprouted tubers in 0.25% Mancozeb or 0.2% Metalaxyl for 30 min.
- Mancozeb (0.25%) or Chlorothalonil (0.2%) or Metalaxyl + Mancozeb (0.2%) or BM (1%) can be applied at 7 to 10 days intervals in the hills and 10 to 15 days intervals in plains.

2. EARLY BLIGHT

Alternaria solani

Economic Importance

- The early blight is a common occurrence wherever potatoes are grown in the world.
- It occurs in cool as well as warm areas.
- Thus, in India the disease is equally serious on the hills as well as in the plains.

Symptom

- Brown-black necrotic spot-angular, oval shape characterized by concentric rings.
- The concentric ridges produce a target board effect, the most characteristic symptom of the disease.
- Several spot coalesce & spread all over the leaf.
- In severe attacks leaves shrivel and fall down.
- Infection and rotting of tubers has also been reported.

Pathogen

- Hyphae are light brown or olivaceous which become dark coloured with age.
- The hyphae are branched, septate and inter and intra cellular.
- The conidiophores emerge through the stomata or between the epidermal cells.
- The conidia are club shaped with a long beak which is often half the long of the whole conidium.
- The lower part of the conidium is brown while the neck is colorless.

- The body of the conidium is divided by 5 - 10 transverse septa and there may or may not be a few longitudinal septa.

Disease Cycle

- The conidia and the mycelium in the soil or in the debris of the affected plants can remain viable for more than 17 months.
- These conidia or the new conidia found on the overwintered mycelium bring about the primary infection of the succeeding potato crop.
- The conidia formed on the spots developed due to primary infection are disseminated by wind to long distances.
- The conidia from the affected plant may also be disseminated to the adjoining plants by rain and insects.
- Infection occurs through the stomata but direct penetration may also takes place.

Favourable Condition

- Dry warm weather with intermittent rain .
- Poor vigor.
- Temperature: 25-30°C.
- Poorly manured crop.

Management

- Disease free seed tubers should be used for planting.
- Removal and destruction of infected plant debris should be done because the spores lying in the soil are the primary source of infection.
- Very early spraying with Zineb or Mancozeb (0.2%) and repeating it for every 15 - 20 days gives effective control.

3. BLACK SCURF

Rhizoctonia solani (Perfect stage-Thanetophorus cucumeris)

Economic Importance

- Black scurfing results in poor marketability of tubers and consumers acceptance.

- Major reduction in yield is through the stem canker phage.
- Pathogen infection reduces the carbohydrate quality of tubers.

Symptoms

- Black speck, black speck scab, russet scab on tubers.
- At the time of sprouting dark brown colour appear on the eyes.
- Affected xylem tissue causes to wilting of plants.
- Infected tuber contains russeting of the skin.
- Hard dry rot with browning on internal tissue.
- Spongy mass appear on the infected tuber.

Pathogen

- The mycelium is hyaline when young and brown at maturity.
- Hyphae are septate and branched with a characteristic constriction at their junction with the main hyphae.
- The branches arise at a right angle to main axis. Sclerotia are black.
- A basidium bears four sterigmata each with a basidiospore at the end.
- The basidiospores are hyaline, elliptical to obovate and thin walled.
- They are capable of forming secondary basidiospores.

Disease Cycle

- The fungus is capable of leading a saprophytic life on the organic material and can remain viable in the soil for several years.
- The sclerotia on the seed tubers is the principal source of infection of the subsequent crop raised with these tubers.
- On return of favourable conditions the mycelium present in the soil may develop producing new hypae.
- The new hypha infects fresh tubers, germinating buds and stem.

- Under adverse condtion, the mycelium turns into sclerotia and serves as means for survival.

Favourable Condition

- Moderately cool, wet weather.
- Optimum temperature for infection is 18°C.
- The black scurf develp more in sandy soil than in clay.

Management

- Disease free seed tubers alone should be planted.
- Treating seed tubers with mercuric chloride solution for 1.5 hr with acidulated mercuric chloride solution for 5 min.
- Well sporulated tubrs planted shallow.
- The disease severity is reduced when land is left fallow for 2 years.
- Spray Mancozeb (0.2%) or Copper Fungicides.

4. COMMON SCAB

Streptomyces scabies

Economic Importance

- The disease is prevalent in Indo – Gangetic region.
- Scabing on potato skin results in lower market quality.
- The infection leads to disfigure potato ultimately poor customer appeal.

Symptoms

- Small brownish and slightly raised spots on tubers.
- Spots enlarge, coalesce and become corky.
- Lesions typically possess a raised margin and slightly depressed center.
- Characteristic symptoms have descriptive names:
  - Russet scab appears on tubers as superficial tan to brown corky lesions.

- Pitted scab is characterized by lesions with depressions beneath the tuber surface.
- Raised scab appears as cushion like warty lesions.

Pathogen

- The pathogen is a member of actinomycetes.
- The actinomycetes tend to form filamentous bead like growth.
- Reproduction is mainly germination of the aerial spores formed in chain and sometimes by growth of fragements of the substrate vegetative non-septed mycelium.
- These are Gram positive bacteria.
- The growth of the organisam is good in slightly alkaline medium and is checked at pH 5.2.

Disease Cycle

- Primary Infection: Soil and infected tubers.
- Secondary Infection: Soil, water, wind blown soil and infected tubers.
- Pathogen enters through unsuberized lenticels or wounds.
- Pre suberization is the most susceptible stage.

Favourable Condition

- Soil pH (5.2-8.0)
- Soil temperature of 20-22°C
- Low soil moisture or dry soil.
- Wetness suppresses the activity.

Management

- Use of disease free tubers.
- Crop rotation with wheat-oat or potato-onion-maize (4yrs).
- Hold the soil pH at about 5.3 by addition of sulphur.
- Green manuring before planting potato.
- Dipping of infected tubers in 3% boric acid for 30 min.

5. BACTERIAL BROWN ROT

Ralstonia solanacearum

Economic Importance

- Bacterial wilt is one of the most important and wide spread bacterial diseases of crops in tropic, subtropics and warm temperate regions of the world.
- It is a serious disease in many part of India including mid hills, the plateau region and West Bengal.
- The disease has been reported on potato, tomato, egg plants and chillies.
- Losses due to this disease range from 10-70 percent.

Symptoms

- Characteristic symptoms are stunting, yellowing of the foliage, wilting and collapse of the whole plant.
- Browning of xylem tissue.
- Eye buds are black in colour.
- Bacterial ooze coming on infected tuber surface and emits a foul odour.

Pathogen

- Rod shaped and motile by 1-4 flagella.
- Gram negative.
- Colonies are white to brown in colour.

Disease Cycle

- Infected soil and seed tubers form the main source of the primary infection.
- Brown rot affected plant parts decay and release masses of bacteria in the soil where these may remain viable from season to season.
- The bacteria in the soil are disseminated by wind , irrigation water and farm implements from one field to the other.
- The infection usually occurs through wounds in the root system.

Favourable Conditions

- High soil temperature (35-37^{0}C).
- Most rapid development of disease takes place at 37^{0}C.
- Soil moisture >50%.

Management

- Deep summer ploughing
- Use disease free tubers for sowing.
- Treat seed tubers with 0.02% Streptocycline for 30 min giving 4mm deep sharp cut.
- Application of stable bleaching powder @ 12kg/ha.
- Crop rotation with maize or soybean for 3 yrs.

6. WART

Synchytrium endobioticum

Economic Importance

- In India, the disease is restricted to Darjeeling.
- Losses seen in temperate regions.

Symptoms

- Host cells at the point of infection are hypertrophied.
- White granular swellings form on the eyes of tubers.
- White tumour like outgrowths, called as warts, develop on stems and tubers.
- As the crop matures, warts become black and rot.
- Infected auxillary bud or the leaf is transformed into cock's comb like greenish yellow outgrowth.

Pathogen

- The fungus is an obligate parasite.

Disease Cycle

- Primary Infection: Resting sporangia in soil.
- Secondary Infection: Zoospores dispersed by water.

Favourable Conditions

- High soil moisture.
- Optimum temperature of 21°C (12 to 28°C)
- Slightly acidic to neutral pH

Management

- Strict quarantine measures
- Grow resistant varieties such as Kufri Jyoti, Kufri Jeevan and Kufri Muthu *etc.*

7. BLACK LEG (SOFT ROT)

Erwinia caratovora subsp. *caratovora*

Economic Importance

- Soft rot of potato, is a common storage and transit disease.
- Sometimes as many as 40 percent plants may die in a badly affected crop.
- Much more loss occurs when the tubers are dug from these fields and transported to warmer areas.
- During transit, there is substantial loss from rotting of tubers.
- Introduced into India through seed stocks from Italy.

Symptoms

- Disease occurs in two phases – Blackleg of shoots and soft rot of tubers.
- Black lesion appear on the base of the plant .
- Systemic and browning of infected tubers.
- Yellow appearance of the plant.
- Finally the plants wilt and die.

- Lenticels (water soaked brown rot) rot and collapse of tubers.
- Soft, reddish or black ring appear on the infected tuber.

Pathogen

- It is a gram negative rod shaped bacterium with 1 to 6 peritrichous flagella.

Disease Cycle

- Survive in diseased and contaminated tubers.
- Spread through contaminated soil, tubers and maggot flies.

Favourable Conditions

- High humidity (94 - 100%).
- Temperature (21 - 29°C).
- Late blight and *Fusarium* tuber rot infections predispose soft rot.

Management

- Remove all debris from warehouses and disinfect the walls with formaldehyde or copper sulphate.
- Avoid wounding of plants and storage organs.
- Products to be stored should be dry, and the humidity and temperatures of warehouses should be kept low.
- Crop rotation with cereals.
- Crop should be planted in well drained areas and at sufficient distances to allow adequate ventilation.
- Dip cut seed pieces of potato in a solution of Streptocycline (100 ppm) and $CuSO_4$ (40 ppm) for 30 min.

8. VIRAL DISEASES

(a) Mild mosaic/Interveinal mosaic – (*Potato virus* X) *PV X*

Symptoms

- Often referred as latent potato mosaic.
- Light yellow mottling with slight crinkling on potato plants.

- Interveinal necrosis of top foliage.
- Stunting of diseased plants.
- Leaves may appear slightly rugose where strains of PV Y combines.

Spread

- Spreads mechanically through rubbing of leaves, contact of infected plants, seed cutting knives, farm implements.
- Root clubbing of healthy and diseased plants in field.

Management

- Disease free seed tubers for planting.
- Rouging of diseased plants.

(b) Severe mosaic – *Potato virus* Y (*PV Y*)

- Also called **potato leaf drop streak**

Symptoms

- Chlorotic streaks on leaves which become necrotic.
- Necrosis of leaf veins and leaf drop streak.
- Interveinal necrosis and stem/petiole necrosis.
- Plant remain stunted in growth.
- Rugosity and twisting of the leaves occurs in combination with PV X and PV A.

Survival and Spread

- Infected tubers.
- Spread by aphids, *Myzus persicae* and *Aphis gossypii.*

Management

- Disease free seed tubers for planting.
- Rouging of diseased plants.
- Aphid control.

(c) Leaf roll – *Potato leaf roll virus*

Symptoms

- Upward rolling of leaves, which have a stiff leathery texture.
- Plants stunted and have a stiff upright growth.
- Phloem necrosis of tubers in some varieties.

Spread

- Infected seed tubers or by aphids.

Management

- Disease free seed tubers for planting.
- Aphid control.

9. POTATO SPINDLE TUBER

Viroid

Symptoms

- Plants appear erect, spindly and dwarfed.
- Leaves small, erect and leaflets dark green.
- Tubers elongated with tapering ends.
- Tuber eyes are numerous and more conspicuous.

Spread

- Infected seed tubers.
- Mechanically spread by knives used to cut seed tubers.
- Also transmitted by pollen and seed and contaminated mouth parts of grasshoppers, flea beetles and bugs.

Management

- Use of PSTVd free potato seed tubers.
- Disinfestation of cutting knive.

Early blight

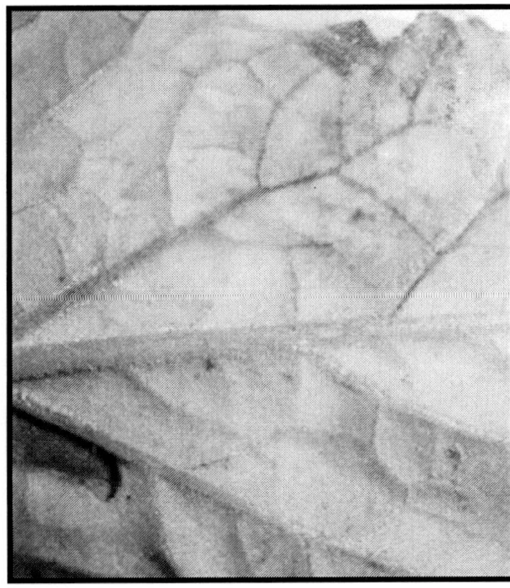

Late blight

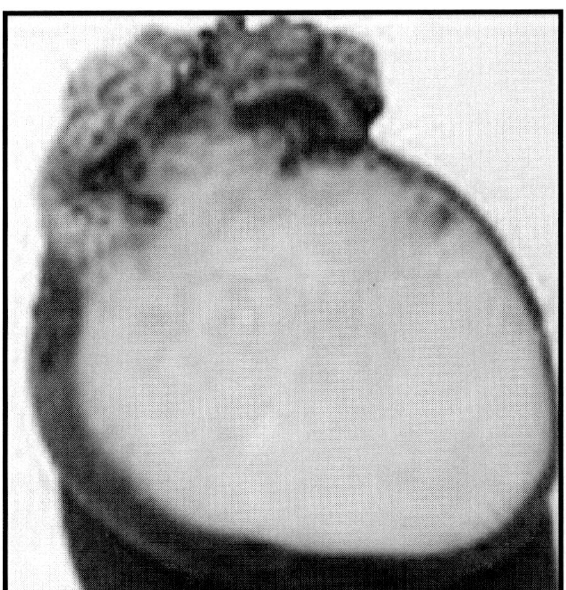

Wart

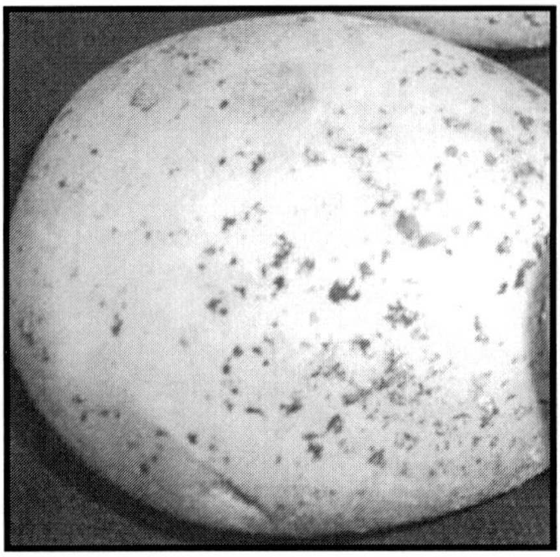

Black Scurf

Black leg/Soft rot

CHAPTER - 15

Diseases of Tomato and Their Management

SI.No	Major diseases	Causal organism
1.	Damping Off	*Pythium aphanidermatum*
2.	Early Blight	*Alternaria solani*
3.	Fusarium wilt	*Fusarium oxysporum* f. sp. *lycopersici*
4.	Bacterial wilt	*Burkholderia solanacearum*
5.	Mosaic	Tomato Mosaic Virus (TMV)
6.	Leaf Curl	Tomato Leaf Curl Virus (TLCV)
7.	Root knot nematode	*Meloidogyne javanica*

1. DAMPING OFF

Pythium aphanidermatum

Economic Importance

- Damping off is one of the common diseases of tomato, and also of a few other vegetable grown from transplanting or even directly planted.
- This is particularly severe in nursery where the seedlings are grown crowded.
- The disease is responsible for poor germination and stand of seedlings in nursery beds.
- Often, the infected seedlings carry the pathogens to the main field where transplanting is done.

Symptoms

- Disease occurs in two stages, *i.e.* the pre-emergence and the post-emergence phase.
- In the pre-emergence phase, the seedlings are killed just before they reach the soil surface.
- The young radical and the plumule are killed and there is complete rotting of the seedlings.
- The post-emergence phase is characterized by the infection of the young, juvenile tissues of the collar at the ground level.
- The infected tissues become soft and water soaked. The seedlings topple over or collapse.

Pathogen

- The fungus has a characteristic mycelium, which coenocytic, hyaline, freely branching and thick hyphae.
- It reproduces both asexually and sexually.
- The asexual reproduction is through the formation of sporangia.
- The sporangium germinates by giving rise to a tubular structure which ends in a vesicle.
- The zoospores are kidney shaped with two flagella.
- The zoospores germinate to produce a germ tube which gives rise to a characteristic coenocytic mycelium.
- The sexual reproduction of the fungus is characterized by the formation of oogonium, antheridium and oospore.
- Oospore is aplerotic, i.e does not fill the oogonial cell completely.
- The oospores may germinate when conditions are favourable, giving rise to germ tube and mycelium.

Disease Cycle

- The fungus is mainly soil borne.
- The oospores persist in soil, resisting adverse conditions.
- When there is sufficient moisture, they germinate and produce the mycelium which infect the host plant.

- The mycelia stage of the fungus is capable of infection the host plant and multiplying very rapidly.
- The zoospores formed in the vesicles of sporangia are commonly released in soil water to swim about and spread from place to place.
- When sexual reproduction starts, resulting in the formation of oospores which help to survive adverse environmental conditions.
- Thus the fungus is capable of living for many years in soil, completing its life cycle both saprophytically and as a facultative parasite.

Management

- Used raised seed bed
- Provide light, but frequent irrigation
- Provide better drainage system.
- Drench with Copper oxychloride (0.3%) or Bordeaux mixture (1%).
- Seed treatment with fungal culture *Trichoderma viride* (4 g/kg of seed) or Thiram (3 g/kg of seed).
- Spray 0.2% Metalaxyl, when there is cloudy weather.

2. EARLY BLIGHT

Alternaria solani

Economic Importance

- Early blight is a common disease wherever tomato is grown.
- Under severe outbreak losses upto 70 percent are common.
- In india, the disease is severe in june to july sown crop than in the winter crop.

Symptoms

- The disease occurring on the foliage at any stage of the growth.
- Early blight is first observed on the plants as small, black lesions mostly on the older foliage.

- Spots enlarge, and by the time they are one-fourth inch in diameter or larger, concentric rings can be seen in the center of the diseased area.
- Tissue surrounding the spots may turn yellow. If high temperature and humidity occur at this time, much of the foliage is killed.
- Lesions on the stems are similar to those on leaves, sometimes girdling the plant if they occur near the soil line.
- Transplants showing infection by the early blight fungus often die when set in the field. The fungus also infects the fruit, generally through the calyx or stem attachment.
- Lesions attain considerable size, usually involving nearly the entire fruit; concentric rings are also present on the fruit.

Pathogen

- Mycelium is septate, branched, light brown which become darker with age.
- Conidiophores are dark colored.
- Conidia are beaked, muriform, dark colored and borne singly.

Disease Cycle

- The disease spreads through infected seeds.
- It may survive in seed bed soils and field, with plant refuse.
- Pathogen already present in soil and plant debris causes primary infection on next crop by wind, rain, irrigation and water splash on the lower leaves.
- Secondary spread occurs through conidia developed on primary spots.
- Infected seed also produce diseased seedlings and serves as primary source.
- Infection takes place through stomata and wounds.

Environmental Conditions

- Optimum temperature-20-30^{0}C.
- Dry warm weather alternating with the intermittent rains.
- Reduction in plant vigour and senescence.

Management

- Removal and destruction of crop debris.
- Practicing crop rotation helps to minimize the disease incidence.
- Spray the crop with Mancozeb (0.2%) or Chlorothalonil (0.2%) or Zineb (0.25%) at weekly interval.

3. FUSARIUM WILT

Fusarium oxysporum f. sp. *lycopersici*

Economic Importance

- This is a severe disease of tomato, occurring in many parts of the world including India, where it is found in every state.
- In some area where tomato is grown repeatedly in the same soil, the disease is severe.

Symptoms

- The first symptom of the disease is clearing of the veinlets and chlorosis of the leaves.
- The younger leaves may die in succession and the entire may wilt and die in a course of few days. Soon the petiole and the leaves droop and wilt.
- In young plants, symptom consists of clearing of veinlet and dropping of petioles. In field, yellowing of the lower leaves first and affected leaflets wilt and die.
- The symptoms continue in subsequent leaves. At later stage, browning of vascular system occurs.
- Plants become stunted and die.

Pathogen

- Mycelium is septate and hyaline.
- They produce macro and micro conidia.
- Micro conidia are one celled, hyaline, ovoid to ellipsoid.
- Chlamydospores are formed in older mycelium.

- Three races of pathogen, Races 1,2, and 3, have been identified.

Disease Cycle

- The fungus is seed borne and soil borne.
- The fungus survives in the soil as chlamydospores or as saprophytically growing mycelium in infected crop debris for more than 10 years.
- One of the chief methods of its distribution is by seedlings raised in infected soil.
- Transfer of conidia through surface drainage water and agricultural implements also help in distribution of the pathogen from field to field.

Environmental conditions

- Optimum temperature-28^{0}C.
- Other conditions which predispose the plant to wilt are short day length, low light intensity, low nitrogen and phosphorus and high potassium.

Management

- Summer ploughing.
- The affected plants should be removed and destroyed.
- Seed treatment with Carbendazim (2.5g/kg seed).
- Spot drench with Carbendazim (0.1%).
- Crop rotation with a non-host crop such as cereals.

4. BACTERIAL WILT

Burkholderia solanacearum

Economic Importance

- The bacterial wilt is a major disease of tomato.
- In Karnataka, Madhya Pradesh, Marathawada region of Maharashtra and West Bengal this disease poses a constant threat to tomato and egg plant.

- The loss in yield may vary between 10 to 90 percent depending on the stage and at which infection occurs.

Symptom

- Characteristic symptoms of bacterial wilt are the rapid and complete wilting of normal grown up plants.
- Lower leaves may drop before wilting. Pathogen is mostly confined to vascular region; in advance cases, it may invade the cortex and pith and cause yellow brown discolouration of tissues.
- Infected plant parts when cut and immersed in clear water, a white streak of bacterial ooze is seen coming out from cut ends.

Pathogen

- The bacterium is gram negative, rod shaped often occurs in pairs, motile with 1 - 4 flagella..

Disease cycle

- The bacterium is soil borne, survives for 3 years in fallow and for a unlimited period in cultivated land.
- Infection always occurs through wounds caused by transplanting, cultural operation or nematode infections.
- The organism first moves into the large xylem vessels.
- When a single lateral bundle is invaded drooping of leaves is common but if all bundles are invaded the plant wilt.
- Chilli, egg plant, grount nut, potato and tobacco are alternative hosts which help it to survive between tomato crops.

Environmental condition

- Relatively high soil moisture and soil temperature favour disease development.
- The optimum temperature for growth is 30 - 37°C.

Management

- Avoid damage to seedling while transplanting.

- Apply bleaching powder @ 10kg/ha.
- Crop rotations, viz., cowpea-maize-cabbage, okra-cowpea-maize, maize- cowpea-maize and finger millet-egg plant are reported effective in reducing bacterial wilt of tomato.

5. MOSAIC

Tomato mosaic virus (TMV)

Economic Importance

- The viral infected plants yield lesser and less number of fruits.
- Sometimes the fruits are deformed.
- Under severe conditions losses upto 50 percent have been observed.

Symptom

- The disease is characterized by light and dark green mottling on the leaves often accompanied by wilting of young leaves in sunny days when plants first become infected.
- The leaflets of affected leaves are usually distorted, puckered and smaller than normal. Sometimes the leaflets become indented resulting in "fern leaf" symptoms.
- The affected plant appears stunted, pale green and spindly.
- The virus is spread by contact with clothes, hand of working labour, touching of infected plants with healthy ones, plant debris and implements.

Pathogen

- Virus particles are rod shaped, not enveloped, usually straight with helical construction and measure 300 x 8nm.
- The single stranded RNA constitutes about 5 percent of the particle weight.
- Thermal inactivation point is 85 - 90°C.

Mode of Spread and Survival

- The virus is seed borne and upto 94% of seeds may contain the virus.

- The virus infection occurs during transplanting It is readily transmissible.
- Many solanaceous plants are susceptible to tomato mosaic virus.
- The virus is spread easily by man and implements in cultural operations or by sap or by animals and by leaf contact.
- The virus particles get entry in the cells after contact and induce symptoms.

Management

- Seeds from disease free healthy plants should be selected for sowing.
- Soaking of the seeds in a solution of Trisodium Phosphate (90 g/litre of water) a day before sowing helps to reduce the disease incidence.
- In the nursery, all the infected plants should be removed carefully and destroyed.
- Seedlings infected with the viral disease should not be used for transplanting.
- Crop rotation with crops other than tobacco, potato, chilli, capsicum, brinjal, etc. should be undertaken.

6. LEAF CURL

Tomato leaf curl virus

Economic Importance

- The leaf curl is a major disease of tomato.
- In India, when the plants are infected within 20 days of planting the losses may be upto 92 percent. While infections of 35 and 50 days old crop result in 74 and 29 percent loss, respectively .

Symptom

- Leaf curl disease is characterized by severe stunting of the plants with downward rolling and crinkling of the leaves. The newly emerging leaves exhibit slight yellow colouration and later they also show curling symptoms.

- Older leaves become leathery and brittle. The nodes and internodes are significantly reduced in size.
- The infected plants look pale and produce more lateral branches giving a bushy appearance. The infected plants remain stunted.

Pathogen

- The particles of the virus have a geminate structure, that is, they are made up of two incomplete icosahedral (isometric particles).
- It is a gemini virus
- The nucleic acid is a circular DNA molecule.
- The virus particles are 80 nm in diameter.

Disease Cycle

- It is neither seed nor sap transmissible.
- But seeds from fresh fruits having infection may have the virus on the seed coat.
- The virus is transmitted by white fly, *Bemisia tabaci* and grafting.
- Even a single viruliferous insect is able to transmit the virus.

Management

- Keep yellow sticky traps @ 12/ha to monitor the white fly.
- Raise barrier crops-cereals around the field.
- Removal of weed host. Protected nursery in net house or green house.
- Spray Imidachloprid (0.05 %) or Dimethoate (0.05%) 15, 25, 45 days after transplanting to control vector.

7. ROOT KNOT NEMATODE

Meloidogyne Javanica

Economic Importance

- Root knot is the most important group of phytonematodes and was first recorded in greenhouse vegetables in England in the year 1855.
- Of the four predominant species of *Meloidogyne, M. incognita* and *M. javanica* are commonly found in the tropics, whereas *M. arenaria* is more common in the sub-tropics.

Symptoms

- The disease is characterized by severe knotting of the roots.
- The first symptoms of the disease is stunting and poor fruiting.
- If the affected plants are pulled out and examined, the roots would be found severely knotted.
- In some cases the plants would show chlorosis but rarely do the plants die from nematode infection.

Disease Cycle

- The nematodes is sedentary endoparasite on roots.
- Second stage juvenile is infective.
- Juveniles and eggs survive in the periods of moisture stress in a state of anhydrobiosis.
- Irrigation water and soil adhering to animal feet and implements help in dissemination of juveniles and eggs of nematodes.

Environmental Conditions

- The length of life cycle and number of generation in one crop is influenced by temperature.
- Optimum temperature-25-30^{0}C
- The life cycle of *M javanica* is completed in 21-25 days at 26-27^{0}C and in 50-60 days at 14-16^{0}C.

- Thus, during winter months in north India the infection of root knot is low because fewer generations are produced during December- January.

Management

- Crop rotation with graminaceous hosts.
- Inclusion of non-preferred hosts like mustard, sesame, maize, wheat, etc., in the cropping system.
- Intercropping of marigold with tomato reduces nematode population.
- Nursery should be raised in nematode free sites or fumigated beds.
- Deep ploughing of infested fields during summer.
- Flooding the field for prolonged periods.
- Use of biocontrol agents like *Paecilomyces lilacinus* (egg parasite).
- Grow resistant varieties like Hissar Lalit and PNR-7.
- Nursery bed treatment with metham sodium (Vapam) @ 25ml/m^2 or Carbofuran @ 0.39g a.i/m^2.
- Bare root dip treatment in EC formulation of systemic pesticides like Dimethoate or Phenemiphos for 6-8 hours before transplantation.
- Integration of three summer ploughings or use of treated nursery with spot application of Carbofuran @ 1 kg a.i/ha at transplanting stage have been found successful to manage nematode population in tomato.

Early blight

Late blight

Damping off

Fusarium Wilt

Mosaic

Leaf Curl

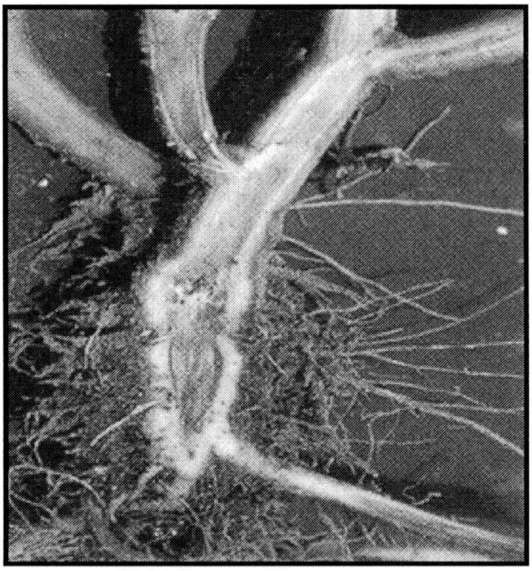

Bacterial Wilt

CHAPTER - 16

Diseases of Brinjal and Their Management

Sl. No	Major diseases	Causal organism
1.	Damping off	*Phythium aphanidermatum*
2.	Phomopsis fruit rot or blight	*Phomopsis vexans*
3.	Collar rot	*Sclerotium rolfsii*
4.	Bacterial Wilt	*Pseudomonas solanacearum*
5.	Little leaf	*Mycoplasma*

1. DAMPING OFF

Pythium aphanidermatum

This disease of brinjal seedlings in nursery beds, has been described under tomato diseases.

2. PHOMOPSIS FRUIT ROT OR BLIGHT

Phomopsis vexans (P.S: *Diaporthe vexans*)

Economic Importance

- The disease was reported in India in 1935 from Gujarat.
- The pathogen attacks foliage and fruits, but the latter phase is more destructive.
- The infection on leaf, stem, seedling and fruit often reduces the production, marketability and seed quality.
- The losses in yield is reported from 20-50 percent.

Symptoms

- The plants are attacked at all stages of growth.
- On leaves, circular to irregular, clearly defined grayish brown spots having light centers appear. The diseased leaves become yellowish in colour and may drop off.
- Several black pycnidia can be seen on older spots.
- The lesions on stem are dark brown, round to oval and have grayish centers where pycnidia develop.At the base of the stem, the fungus causes characteristic constrictions leading to canker development and toppling of plants.
- On fruits, small pale sunken spots appear which on enlargement cover entire fruit surface. These spots become watery leading to soft rot phase of the disease. A large number of dot like pycnidia also develop on such spots.
- The infection of fruit through calyx leads to development of dry rot and fruits appear black and mummified.

Pathogen

- The mycelium of the pathogen is septate and hyaline becoming dark with age.
- Pycnidia are submerged and later becoming erumpent with a prominent ostiole. Conidia are produced on simple to branched conidiophores and are of 2 types: Alpha conidia, which are sub cylindrical and beta conidia, which are filiform and curved.
- Role of beta-conidia in the epidemiology of the disease is not very clear.
- The perfect stage produces perithecia in which asci with 8 hyaline, bicelled, ellipsoid-fusoid ascospores are produced which are usually constricted at septum.

Disease Cycle

- The pathogen is seed borne and also survives in plant debris as mycelium and pycnidia.
- The organism is locally disseminated as water borne pycniospores.

- The spore exude out through the ostiole and can easily be thrown away rain drops.
- They can also disseminated by tools and inscets.
- Inoculum present in the form of mycelium and pycnidia results in the diseased seedlings.

Favourable Conditions

- The fungus requires wet weather and high temperature for best development.
- Maximum disease development takes place at about 26^{0}C under wet weather conditions with 55% relative humidity.

Management

- Use of disease free seed.
- Practicing crop rotation and summer ploughings helps in reducing initial inoculum.
- Hot water treatment of seed at 50°C for 30 minutes.
- Seed treatment with Thiophanate methyl at 1g/kg seed.
- Spray twice with Thiophanate methyl or Carbendazim @ 0.1% at 20 days interval.
- Removal and destruction of diseased crop debris.

3. COLLAR ROT

Sclerotium rolfsii

Economic Importance

- This is a minor disease but many assume serious form if soil conditions are favourable.

Symptoms

- The disease occasionally occurs in serious form.
- The lower portion of the stem is affected from the soil borne inoculum (sclerotia).
- Decortications is the main symptom.

- Exposure and necrosis of underlying tissues may lead to collapse of the plant.
- Near the ground surface on the stem may be seen the mycelia and sclerotia.
- Lack of plant vigour, accumulation of water around the stem, and mechanical injuries help in development of this disease.

Management

- Seed treatment with 4 g of *Trichoderma viride* formulation per kg seed will help in reducing the disease.
- Spraying with Mancozeb @ 2g/litre of water.
- Collection and destruction of diseased parts and portions.

4. BACTERIAL WILT

Pseudomonas solanacearum

Economic Importance

- Bacterial wilt is one of the most widespread and destructive diseases of brinjal.
- It affects a variety of solanaceous crops and causes severe losees upto 75 percent in yield.

Symptoms

- Stunting, yellowing of the foliage and finally sudden collapse of the entire plant are the characteristic symptoms of the disease.
- Lower leaves may droop first before wilting occurs.
- The vascular system becomes brown.
- Bacterial ooze comes out from the affected parts.
- Plant show wilting symptoms at noon time will recover at nights, but die soon.

Pathogen

- The bacterium is non acid fast, non spore forming, non capsulated and motile by a polar flagellum.
- The bacterium produces acid.

Disease Cycle

- The bacterium is both soil and seed borne in nature and overwinters in infected plant parts, in wild host plants and weeds.
- The bacteria spread through irrigation water or infested soil or agricultural implements and enter the plant through wounds.

Favourable Conditions

- Relatively high soil moisture and soil temperature favour the disease development

Management

- Crop rotation with cruciferous vegetables such as cauliflower help in reducing the disease incidence.
- Fields should be kept clean and effected parts are to be collected and burnt.
- Spray Copper fungicides to control the disease (2% Bordeaux mixture).
- Soil solarization with a transparent polyethylene sheet for 8-10 weeks during March-June.
- Biological control with *Pseudomonas fluorescens.*
- Tolerant Variety Pant Smart.

5. LITTLE LEAF

Phytoplasma

Economic Importance

- The disease was first reported from India in 1938 and it is known to oocur only in India and Sri lanka.
- The loss caused by this disease has been reported upto 80 percent.

Symptoms

- Small or little leaves.

- The petioles are so short that the leaves appear to be sticking to the stem.
- Affected plants have narrow, soft, smooth and yellow shorter leaves.
- Newly formed leaves are much shorter.
- Internodes of the stem are also shortened.
- Axillary buds get enlarged but their petioles and leaves also remain shortened giving the plant a bushy appearance.
- Mostly there is no flowering but if flowers are formed they remain green.
- Fruiting is rare, if any fruit is formed, it becomes hard, tough and fails to mature.
- Young fruit turns necrotic, get mummified and cling to the plant.

Pathogen

- The disease is caused by Phytoplasma having ovoid to spherical body which is concentrated in the phloem sieve tubes.

Disease Cycle

- Natural transmission of little leaf disease of brinjal is through the vector *Cestius phycitis* (*Eutetics phycitis*).
- The disease can also transmitted by grafting and by leaf hopper vector *Hishimonas phycitis.*
- Collateral hosts: *Datura fastuosa, D. stramonium, Vinca rosea, Argemone mexicana,* chilli, tomato and tobacco.
- The pathogen perenneates in weed hosts.

Management

- Tolerant variety: Pusa Purple Round, Pusa purple cluster and Arka sheel.
- Destruction of affected plants.
- Eradication of solanaceous weed hosts.
- Spray Malathion 3 ml/lit. or soil application of Phorate.

Phomopsis blight

Little leaf of brinjal

Bacterial wilt

Collar rot

CHAPTER - 17

Diseases of Chilli and Their Management

SI.no	Major diseases	Pathogen
1.	Damping off	*Pythium aphanidermatum*
2.	Fruit Rot or Die back	*Colletotrichum capsici*
3.	Powdery mildew	*Leveillula taurica*
4.	Bacterial leaf spot	*Xanthomonas campestris pv. vesicatoria*
5.	Cercospora leaf spot	*Cercospora capsici*
6.	Fusarium wilt	*Fusarium oxysporum f.*sp.*capsici*
7.	Viral diseases	
a.	Mosaic	Chilli mosaic virus
b.	Leaf curl	Tobacco leaf curl virus

1. DAMPING OFF

Pythium aphanidermatum

Economic Importance

- Damping off is one of the common diseases of chill, and also of a few other vegetable grown from transplanting or even directly planted.
- This is particularly severe in nursery .
- The disease is responsible for poor germination and stand of seedlings in nursery beds.

- Often, the infected seedlings carry the pathogens to the main field where transplanting is done.

Symptoms

- Pre-emergence damping off: Seedlings disintegrate before they come out of soil surface. This is known as pre-emergence damping-off which results in poor field emergence/poor seed germination.
- Post-emergence damping off is characterised by development of disease after seedlings have emerged out of soil surface but before the stems are lignified.
- Lesion formation at collar region.
- Infected areas appear brown and water soaked.
- Plants shrivel and collapse as a result of softening of tissues.
- Infected stems become hard, thin (Wire stem symptoms) and infected seedlings topple.
- Disease appear in patches both in nursery and field beds.

Pathogen

- Mycelium is hyaline, coenocytic and zoosporangia are lobed and branched.
- Zoospores are biflagellate and oogonia are spherical with smooth walled.
- Antheridia are monoclinous, intercalary or terminal.
- Oospores are aplerotic, single with thick wall.

Disease Cycle

- The pathogen is soil borne.
- Zoospores spread through irrigation water.
- The disease spreads to main field by planting infected seedlings.

Favourable Conditions

- Heavy rainfall.
- Excessive and frequent irrigation.

- Poorly drained soil and close spacing.
- High soil moisture with temperature around 25-30^0 C.

Management

- Raise nursery in light soil with proper drainage.
- Sterilization of nursery bed soil with chloropicrin or methyl bromide one week before sowing.
- Soil solarization of nursery beds.
- Sowing seed on raised beds of 6-8" high (15 cm)
- Soil drenching with Copper oxychloride (0.25%) or 1% Bordeaux mixture or Metalaxyl (0.2%).
- Seed dressing with Thiram or Captan @ 2-3 g/kg.
- Biocontrol with *Trichoderma viride* and *Trichoderma harzianum.*

2. FRUIT ROT OR DIE BACK

Colletotrichum capsici

Economic Importance

- Anthracnose, die back, and ripe fruit rot is the most important fungal disease of the chilli crop.
- It is present throughout India but is more common and aggressive in Assam, North Bihar, Andhra Pradesh and parts of Uttar Pradesh.
- Anthracnose is limiting the profitable cultivation and seed production through out the major chilli growing regions of India.
- In favorable weather, it may cause 12 to 25 percent loss in the crop.

Symptoms

The disease occurs in two forms: the die back and the ripe -rot.

Die Back Symptoms

- Small, circular to irregular, brownish black scattered spots appear on leaves.
- Severely infected leaves defoliate.

- Infection of growing tips leads to necrosis of branches from tip backwards.
- Necrotic tissues appear grayish white with black dot like acervuli in the center.
- Shedding of flowers due to the infection at pedicel and tips of branches.

Fruit Symptoms

- Ripe fruits are more liable for attack than the green ones.
- Small, circular, yellowish to pinkish sunken spots appear on fruits.
- Spots increase along fruit length attaining elliptical shape.
- Severe infection result in the shrivelling and drying of fruits.
- Such fruits become white or greyish in colour and lose their pungency.
- On the surface of the lesions minute black dot like fruiting bodies called 'acervuli'.
- Development of concentric rings and fruits appear straw coloured.
- The affected fruits may fall off subsequently.

Pathogen

- The mycelium is septate and inter and intra cellular.
- Conidia in mass appear pinkish.
- Conidia are falcate, fusiform with acute apices and narrow truncated base.
- Acervuli formed are rounded, elongated.
- Setae are brown 1-5 septate, rigid, hardly swollen at the base, slightly tapered towards the paler acute apex.

Disease Cycle

- The pathogen survives in the plant debris, which include infected twigs, stem and fruits.
- The secondary spread take place through the wind borne conidia produced in the infected left out plants.

- The conidia germinate within four hours in water and an appresoria is developed for further growth of the fungus between the plant tissues.
- Infected seeds or badly infected fruits also carry the pathogen and serves as source of primary inoculums.

Favourable Conditions

- Temperature, 28^{0}C with Relative humidity more than 97%.
- Humid weather with rainfall at frequent intervals.

Management

- Collect and destroy all infected plant parts.
- Collect seeds only form fruits without infection.
- Removal and destruction of Solanaceous weed hosts and infected plant debris.
- Resistant varieties: Perennial, Bengal Green, S20-1, H-4, H-6, Lorai, etc.
- Seed treatment with Thiram or Captan or Mancozeb 2.5g/ kg is found to be effective in eliminating the seed-borne inoculum.
- Three sprayings with Mancozeb (0.25%) or copper oxychloride (0.25%) not only reduced the disease incidence but also increased the yield of fruits.

3. POWDERY MILDEW

Leveillula taurica

Economic Importance

- The powdery mildew is one of the important fungal diseases of chillies.
- The disease assumes a serious proportion under dry weather conditions.

Symptoms

- White powdery coating appears mostly on the lower surface.

- Sometimes the powdery coating can also be seen on the upper surface.
- Correspondingly on the upper surface yellow patches are seen.
- Severe infection results in the drying and shedding of affected leaves.
- Powdery growth can also be seen on young fruits, and branches.
- Diseased fruits do not grow further and may drop down.

Disease Cycle

- The disease cycle starts when conidia land on a chilli leaf.
- Conidia germinate much like a seed and begin to grow into the leaf.
- The fungus initially grows unseen within the leaf for a latency period of 18-21 days.
- Then the fungus grows out of the breathing pores (stomates) on the under surface of the leaf, producing spores which are borne singly on numerous, fine strands or stalks .
- These fungal strands become visible as white patches or mildew colonies on the under side of the leaf.
- Repeated cycles of powdery mildew can lead to severe outbreaks of powdery mildew that economically damage the crop.

Favourable Conditions

- Cool dry weather favours conidial germination.
- High relative humidity favours disease development.

Management

- Spray Wettable Sulphur (0.25%) or Karathane (0.05%) or Carbendazim (0.1%).

4. BACTERIAL LEAF SPOT

Xanthomonas campestris pv. vesicatoria

Economic Importance

- The disease appears on leaves and also on green fruits but it is more common as leaf spots.
- The loss caused by this disease is upto 16 percent.
- Three races of the pathogen *X campestris pv. vesicatoria* have reported.

Symptoms

- The disease appears as small, circular to irregular water soaked areas or spots on the lower leaves.
- Spots become necrotic with brown center with chlorotic borders.
- Enlarged spots may develop straw coloured centres.
- Severely spotted leaves turn yellow and drop.
- Raised brown lesions appear on fruits.
- Narrow elongated lesions or streaks may develop on stems.

Disease Cycle

- The disease is primarily seed borne and also survives in infected plant debris.
- It spreads in the nursery and is further disseminated with infected transplants.
- Spattering rains are the chief means of dissemination.
- The bacterium subsists in infected debris.

Favourable Conditions

- The pathogens requires a temperature of 22-34^{0}C with high humidity for maximum infection.
- The disease becomes more prominent after rains.

Management

- Field sanitation.
- Seeds must be obtained from disease free plants.
- Seed treatment with Copper oxychloride 3g/kg seed.

- Seedlings may be sprayed with Bordeaux mixture (1%) or Copper oxychloride (0.3%).
- Spray twice with Agrimycin (100 ppm) or Plantomycin (200ppm) along with 3g Copper oxy chloride per litre of water at 15 days interval.

5. CERCOSPORA LEAF SPOT

Cercospora capsici

Symptoms

- Circular spots with brown margins appear on leaves.
- The spots enlarge and coalesce with others.
- The central portion of the spot becomes white and the leaves turn yellow and defoliate.
- Sometimes central portion of spot drops off.
- Spots also appear on stems and twigs as dark brown, irregular lesions with whitish centers.
- In severe cases die-back of twigs occur.

Pathogen

- Stromata are well developed.
- Conidiophores are 30- 60 x 4.5 – 5.5 micron meter in size.
- Conidia are subhyaline to coloured, circular to obculate.

Disease Cycle

- **Primary Infection:** Dormant mycelium in infected plant debris, infected seeds and volunteer plants.
- **Secondary Infection:** Wind borne conidia.

Management

- Spray twice at 15 days interval with Mancozeb (0.25%) or Copper oxychloride (0.3%).

6. FUSARIUM WILT

Fusarium oxysporum f.sp.capsici

Symptoms

- Characterised by wilting of the plant and upward and inward rolling of the leaves. The leaves turn yellow and die.
- Commonly appear in localised areas of the field where a high percentage of the plants wilt and die, although scattered wilted plants may also occur.
- Disease symptoms are characterised by an initial slight yellowing of the foliage and wilting of the upper leaves that progress in a few days into a permanent wilt with the leaves still attached.

Pathogen

- Mycelium is grayish white.
- Microconidia are formed singly, hyaline and cylindrical.
- Macroconidia are cylindrical to falcate.
- Chlamydospores are globose to oval and rough walled. discoloured, particularly in the lower stem and roots.

Disease Cycle

- The fungus is mainly soil borne and survive in soil mainly as chlaymydospores formed by the hyphal and conidial cells.
- Infection is always through injured roots.
- Deep wounds to expose the xylem help in easy infection.
- After entry into the root, the fungus multiply extensively in vascular tissues, interrupting the transport of nutrients to the above ground plant parts, where the mycelium is formed in abundance to partially or completely plug the vessels, interrupting the transport of nutrients to the above ground plant parts, thus stunting the plant growth.
- Continued dislocation of flow of plant nutrients leads to plant wilt.

Environmental Conditions

- Optimum soil temperature-20-30^{0}C
- Hot and dry summers followed by rains seems to favor the disease.

Management

- Use of wilt resistant varieties.
- Drenching with 1% Bordeaux mixture or Blue copper or Fytolan (0.25%) may give protection.
- Seed treatment with 4g *Trichoderma viride* formulation or 2g Carbendazim per kg seed is effective.
- Mix 2kg *T.viride* formulation mixed with 50 kg FYM, sprinkle water and cover with a thin polythene sheet. When mycelia growth is visible on the heap after 15 days, apply the mixture in rows of chilli in an area of one acre.

7. VIRAL DISEASES

a. **Chilli mosaic virus**

Economic Importance

- Chilli mosaic virus is reported to cause a loss of 20-75 percent in yield.

Symptoms

- Light green and dark green patches on the leaves.
- These may be sunken or raised (puckering).
- The leaves are generally reduced in size and some what become filamentous like.
- Diseased plants produce less flowers and fruits.

Transmission

- The virus is sap transmissible and can be easily transmitted through contact during cultivation.
- Natural transmission of virus through aphid vectors such as *Myzus persicae, Aphis gossypii.*

b. **Tobacco leaf curl virus**

Economic Importance

- The leaf curl is a major limitation in successful cultivation of both hot and bell peppers.
- The loss caused by the disease is reported 100 percent.

Symptoms

- Leaves curl towards midrib and become deformed.
- Stunted plant growth due to shortened internodes and leaves greatly reduced in size.
- Flower buds abscise before attaining full size and anthers do not contain pollen grains.
- The virus is generally transmitted by whitefly. So control measures of whitefly in this regard would be helpful.

Pathogen

- The tobacco leaf curl virus causes the leaf curl disease.

Transmission

- The virus is not sap transmitted and is also not seed borne.
- In nature, it is transmitted by the white fly (*Bemesia tabaci*).

Management of Viral Diseases

- Selection of healthy and disease free seed.
- Nursery beds should be covered with nylon net or straw to protect the seedlings from viral infection.
- Raise 2-3 rows of maize or sorghum as border crop to restrict the spread of aphid vectors.
- Apply Carbofuran 3G @ 4-5 Kg/acre in the main field to control sucking complex and insect vectors selectively.
- If it is not possible spray the crop with systemic insecticides like Dimethoate (2ml) or Monocrotophos (1.5 ml) per litre of water.
- Collect and destroy infected virus plants as soon as they are noticed.

Damping off

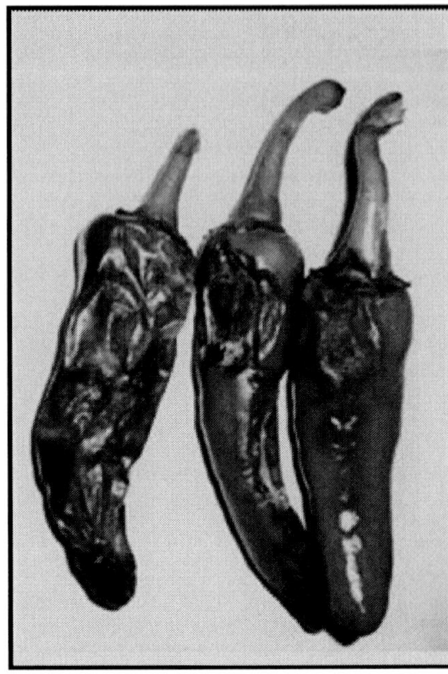

Fruit rot

Powdery mildew

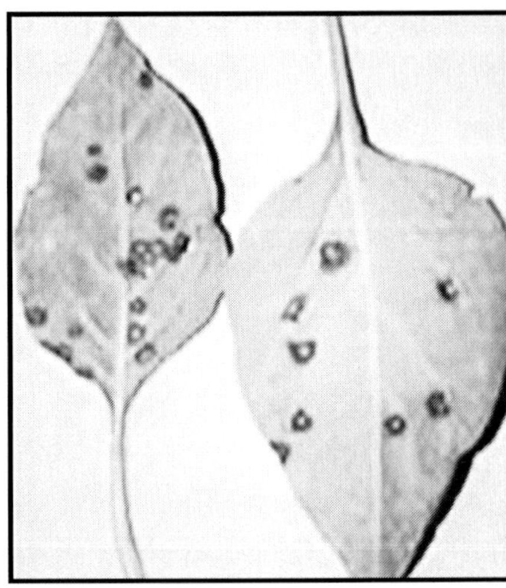

Cercospora leaf spot

Mosaic

Leaf Curl

Fusarium Wilt

CHAPTER - 18

Diseases of Bhendi and Their Management

SI. No	Major diseases	Causal organism
1.	Yellow Vein Mosaic	Bhendi yellow vein mosaic virus
2.	Cercospora Leaf Spot	*Cercospora malayensis, C. abelmoschi*
3.	Wilt	*Fusarium oxysporum f.*sp. *vasinfectum*
4.	Powdery Mildew	*Erysiphe cichoracearum*

1. YELLOW VEIN MOSAIC

Bhendi yellow vein mosaic virus

Economic Importance

- This is a severe disease of the crop all over the India, and is the main limiting factor in the cultivation of the crop extensively.
- The damage caused may be anywhere from 50-100 percent of fruit yield.

Symptoms

- Yellowing of the entire network of veins in the leaf blade is the characteristic symptom.
- In severe infections the younger leaves turn yellow, become reduced in size and the plant is highly stunted.

- The veins of the leaves will be cleared by the virus and intervenal area becomes completely yellow or white.
- In a field, most of the plants may be diseased and the infection may start at any stage of plant growth.
- Infection restricts flowering and fruits, if formed, may be smaller and harder.
- The affected plants produce fruits with yellow or white colour and they are not fit for marketing.
- The virus is spread by whitefly.

Disease Cycle

- **Primary Infection:** Infected plant parts.
- **Secondary Infection:** The virus is transmitted by the whitefly, *Bemisia tabaci.*

Management

- Resistant varieties like Pusa Savani, Parbhani Kranti, Arka Abhay, Arka Anamika, and Varsha Uphar.
- For sowing during the summer season, when the whitefly activity is high, the susceptible varieties should be avoided.
- Spraying Monocrotophos 1.5 ml/litre or Chlorpyriphos 2.5 ml + neem oil 2 ml/lit of water can restrict the disease spread.

2. CERCOSPORA LEAF SPOT

Cercospora malayensis, C. abelmoschi

Symptoms

- In India, two species of *Cercospora* produce leaf spots in bhendi.
- C. *malayensis* causes brown, irregular spots and
- *C. abelmoschi* causes sooty black, angular spots.
- Both the leaf spots cause severe defoliation and are common during humid seasons.

Pathogen

- Conidiophores are pale to medium olivaceous brown, multiseptate, some times branched, geniculate and irregular.

- Conidia are obclavate to cylindric, olivaceous brown and straight to curved.

Disease Cycle

- The fungus survive through conidia and stomata on crop debris.
- The germ tubes from conidia enter the host through stomata and causes new infection.

Favorable Conditions

- These species causes maximum infection at 25-29^{0}C.

Management

- Spraying Bordeaux mixture (1%), Carbendazim (0.1%), Mancozeb (0.25%) control the disease.

3. FUSARIUM WILT

Fusarium oxysporum f. sp. *vasinfectum*

Economic Importance

- This disease had not been of any significance in India however, of late reports of its occurrence with some severity have appeared.
- The disease is found wherever okra is grown intensively.

Symptoms

- The noticeable symptom is a typical wilt, beginning with a yellowing and stunting of the plant, followed by wilting and rolling of the leaves as if the roots were unable to supply sufficient water.
- Finally, the plant dies.
- If the diseased stem is slpit length wise the vascular bundles appear as dark streak.
- When severely infected, nearly the whole stem is blackened

Pathogen

- The mycelium is hyaline and intracellular in the host.
- Macroconidia are 3-5 septate formed on sporodochia.
- They are fusiform - falcate in shape and curved inward at both ends.
- Microconidia are septate (Zero to one septa).
- Terminal and intercalary chlamydospores are formed.

Disease Cycle

- The fungus is soil borne.
- The fungus survives in the soil as chlamydospores or as saprophytically growing mycelium in infected crop debris.
- In contact with host roots the chlamydospores or conidia germinate and penetrate the roots.
- After some growth in the cortex the fungus reaches the xylem where it multiplies very rapidly.
- Dissemination of the fungus can occur by any method that can transfer the soil from one place to another.

Favorable Conditions

- The optimum temperature for growth of the fungus is 25^0C and for disease development 22^0-28^0C.

Management

- Treat the seeds with Mancozeb @ 3g/kg seed. Drench the field with Copper oxychloride @ 0.25%.

4. POWDERY MILDEW

Erysiphe cichoracearum

Economic Importance

- The disease has been recognized as a major disease both under field and green house conditions, the world over.
- Powdery mildew may result into considerable yield losses under dry weather conditions.

Symptoms

- The fungus produces greyish powdery growth on the under as well as on the upper surface of the leaf causing severe reduction in fruit yield.

Pathogen

- Conidia are single celled, hyaline, barrel shaped and in long chains.
- Cleistothecia are globose and dark brown myceloid appendages.
- The asci are pedicellate, ovate or ellipsoid.
- The number of ascospores is usually 2 rarely 3 per ascus.
- The ascospores are single celled, hyaline and oval to sub cylindrical.

Disease Cycle

- The fungus can overwinter in the form of mycelium and conidia on different cucurbitaceous hosts and this may serve as means of primary inoculums.
- Wild cucurbits harbour the conidial stage of the fungus and release conidia for primary infection to the spring or summer sown cucurbits.
- Secondary infection occurs through conidia spread by wind, thrips and other insects.

Management

- Spary inorganic Sulphur (0.25%) or Karathane (0.1%), Carbendazim (0.1%) 3 or 4 times at 15 days interval.

Yellow vein mosaic

Powdery mildew

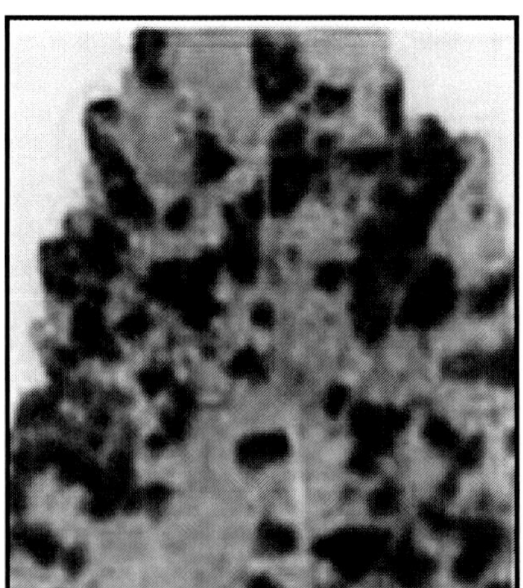

Cercospora leaf spot

Chapter - 19

Diseases of Crucifers and Their Management

Sl. No	Major diseases	Causal organism
1.	Club root or Finger and toe	*Plasmodiophora brassicae*
2.	Downy mildew	*Peronospora parasitica*
3.	Powdery mildew	*Erysiphe polygoni*
4.	Alternaria leaf spot	*Alternaria brassicola, A.brassicae*
5..	White rust	*Albugo candida (Syn: Cystopus candidus)*
6.	Black rot	*Xanthomonas campestris pv. campestris*
7.	Bacterial soft rot	*Erwinia carotovora*

1. CLUB ROOT OR FINGER AND TOE

Plasmodiophora brassicae

Economic Importance

- The disease attacks a wide range of plants such as cabbage, cauliflower and other crucifers.
- It is widely distributed all over the world, but occurs most frequently in Europe and North America.
- In India, the disease has been reported occurring generally in restricted hilly areas of West Bengal and Assam.
- Fields once infested with the club root pathogen remain so indefinitely and become unfit for cultivation of crucifers.

Symptoms

- Characteristic symptoms become apparent in advanced stage of root infection as spindle like, spherical, knobby, or club-shaped swellings.
- Stunting and yellowing of plants.
- Leaves become yellowish and wilt on hot days.

Pathogen

- The pathogen is an obligate endoparasite inhabiting the host cells in the form of plasmodium.
- The zoospores are naked, devoid of a cell wall and amoeboid in shape.
- The zoospore is biflagellate and one of the flagella is much longer than other.

Disease Cycle

- The fungus persist in soil as resting spore in diseased plant tissue.
- When a susceptible host is planted in the soil it infects the root, causing club root symptoms.
- It multiplies and spreads in the plasmodial stage, causing hyperplasia of the host cells and later forms the resting spores.
- Thus, the disease cycle is completed.

Favourable Conditions

- It occur at a temp. range of 20-25 ^{0}C.
- High soil moisture (45-60%).
- Neutral to acidic soils (5-7.0 pH).

Management

- Use of seedlings from disease free fields.
- Plant cabbage and other susceptible cruciferous crops in well drained fields that have a pH slightly above neutral (usually about pH 7.2).

- Crop rotation does not work since pathogen persists long in soil.
- Add hydrated lime to soil to increase pH to 7.2 (6 weeks before planting @ 2.5ton/ha).
- Avoid excess irrigation.
- Soil fumigation with Methly bromide, 1kg/10m^2 followed by covering with plastic film.
- Seed treatment with Captan or Thiram (4g/kg), followed by *T.viride* 4g/kg seed.

2. DOWNY MILDEW

Peronospora parasitica

Economic Importance

- The disease is prevalent in all the cooler parts of the world.
- The cruciferous crops especially cabbage and cauliflower grown in hills during summer season are badly damaged by this disease.
- The infection is commonly among young plants but does not progress much after the plants attain a height of more than half feet.

Symptoms

- It may attack young plants and also at the seed production stage as being commonly observed in northern India in recent years, when high humidity prevails during seed production stage.
- The fungus when attacks the young seedlings, discoloration occurs and in severe cases the whole plant perishes.
- Purplish leaf spots or yellow brown spots on the upper surface of the leaf appear, while fluffy downy fungus growth is found on the lower surface.

Pathogen

- The pathogen is an obligate parasite.
- Its hyphae are coenocytic, strictly intercellular with finger shaped, branched haustoria.

- The conidiophores arise vertically and their base is flattened and dichotomously branched.
- The conidia are broadly oval, ellipsoidal and hyaline, borne singly at the tip of each branch.
- The oospores are globose and appear pale yellow in color.They germinate by germ tube.

Disease Cycle

- The pathogen penetrates in the soil through its oospores, which are brought therein along with infected plant debris.
- The oospores perennating in soil germinate by a germ tube at the return of the favourable conditions.
- The germ tube infects underground parts of young seedlings causing primary infection.
- The secondary infection takes place by wind borne condia.They are brought on to the susceptible host, germinate by germ tube , which penetrates the host and establishes the secondary infection in the growing season.
- During the end of the growing season the pathogen enters sexual phase of life cycle and produces oospores in the infected host parts.
- These oospores, however perennates and serve as the source of primary infection during the next growing season.

Management

- Destruction of infected plant debris.
- Avoidance of thick sowing and excessive moist conditions.
- Spray Metalaxyl (0.25%) or Copper Oxy Chloride (0.3%) or Mancozeb (0.2%) at 15 days interval.

3. POWDERY MILDEW

Erysiphe polygoni

Symptoms

- White powdery spots on the upper surface of leaves, stems, flower parts etc.
- Finally the mildew may cover the entire surface.

Pathogen

- Conidiophores are septate.
- The cleistothecia are sharp and globose.

Disease Cycle

- Primary Infection: Mycelium in infected plant debris
- Secondary Infection: Wind borne conidia.

Management

- Spary inorganic Sulphur (0.25%) or Karathane (0.05%).

4. ALTERNARIA LEAF SPOT

Alternaria brassicola, A.brassicae,

Economic Importance

- Black leaf spot or *Alternaria* leaf spot is one of the limiting factor in the production of cruciferous vegetables especially raised for seed purpose.
- The disease is prevalent all over the world.
- It is of concern in north and central India where appreciable seed losses in terms of size, color and germination inability are reported.

Symptoms

- Spots are small, dark coloured.
- They enlarge, soon become circular & 1mm. in diameter.
- Under humid conditions groups of conidiophores will be formed in the spot.
- Spots develop concentric rings.
- Finally, the spots coalesce leading to blighting of leaves.
- The fungus is seed borne and cause shriveling of seeds and poor germination.
- Linear spots also appear on petioles, stems, pods & seeds.

Pathogen

- The fungal hyphae are branched, septate, inter and intracellular.
- Conidiophores arise singly or in groups of 2 to 12.
- They are simple, erect, cylindrical, slightly swollen at base, septate, pale, smooth and 90 x 5 to 8 mm. Conidia are formed in chains of 20 or more.
- They are cylindrical, muriform, tapering slightly towards the apex and the basal cell is rounded.

Disease Cycle

- Primary Infection: Mycelium persisting in the seed or as spores on seed or from debris.
- Secondary Infection: Wind or insect borne conidia.
- On infection, pathogen becomes subcuticular in leaves. This is followed by colonization of epidermal and mesophyll cells.

Environmental Conditions

- Optimum temperature- 16-26^{0}C
- Relative humidity-79-96%
- Wind velocity-2.5 to 6 km/hr

Management

- Hot water treatment at 50°C for 30min.
- Seed treatment-Thiram or Captan @ 2.5g/kg seed.
- Foliar spray with Mancozeb (0.2%) or Copper Oxy Chloride (0.3%) twice.

5. WHITE RUST

Albugo candida (Syn: Cystopus candidus)

Economic Importance

- White rust is quite serious on cruciferous seed crops of India.
- The yield losses are generally substantial where white rust and downy mildew occur together.

Symptoms

- White, shiny raised blisters (pustules) on the lower surfaces of leaves, stems & flowers.
- Pustules coalesce to form irregular patches.
- The epidermis ruptures exposing white spore mass which gives the pustule a powdery appearance.
- Distortion of the floral parts including petals, pistils and anthers due to hypertrophy and hyperplasia.
- Plants malformed beyond recognition.

Pathogen

- The Pathogen is an obligate parasite.
- Mycelium is intercellular producing knob shaped haustoria in the host cells.
- The conidia are spherical, thick walled, hyaline and formed in chains.
- The oospore is spherical, with a thick irregular wall, deep yellow to dark brown in color.
- The oospore germinate to produce a vesicle, inside with zoospores are formed.
- The zoospore are elliptical to kidney shaped and are biflagellate.
- They soon lose the flagella , become spherical and then germinate by a germ tube, which produces a mycelium.

Disease Cycle

- Primary Infection: Oospores in soil and Perennial weeds hosts.
- Secondary Infection: Sporangia & Oospores carried by wind.
- The sporangia germinate to release zoospores which infect the host through stomata.

Environmental Factors

- Optimum temperature -15-20^{0}C
- Relative humidity-Low (40%)
- Weather- cool and moist

Management

- Sanitary measures.
- Crop rotation.
- Destruction of weeds.
- Spray 0.8% B.M or any copper fungicide.

6. BLACK ROT

Xanthomonas campestris pv. campestris

Economic Importance

- Black rot occurs world wide on many cultivated and wild crucifers and is particularly destructive in subtropical or warm temperate regions.

Symptoms

- First appear near the leaf margins as chlorotic or yellow (angular) areas.
- The yellow area extends to veins & mid rib forming characteristic 'v' shaped chlorotic spots.
- Veins and veinlets turn brown and finally black.
- The vascular blackening extend beyond affected veins to midrib, petiole and stem.
- In advanced stages, infection may reach the roots system and blackening of vascular bundles occur.
- Bacterial ooze can also be seen on affected parts.
- If the infection is early, the plants wilt and die. If the infection is late plant succumb to soft rot & die.

Pathogen

- It is gram negative, short rod with rounded ends and non capsulated.
- It occurs singly, rarely in pairs and motile with single polar flagellum.

Disease Cycle

- Bacterium is internally seed and soil borne. It also survives on plant debris.
- Geminating infected seeds serve as the primary source of inoculums.
- Bacteria enter the cotyledons through stomata at the marginal sinus and pass from the cotyledons of young leaves and progress systematically throughout the plant.
- The foliage infection through water pores and insect injuries.
- The chief source of local transmission in the field is by irrigation water or wind splashed rain, cultural practices and transplant.

Environmental factors

- Black rot is spread rapidly during warm, humid weather.
- Optimum temperature - 27- 30°C
- Relative humidity-80- 100%

Management

- Seed treatment with Aureomycin (1000 ppm) for 30 min.
- Hot water treatment at 50°C for 30 min, for killing seed borne inoculum followed by a 30 min dip in Streptocycline (100ppm).
- Drenching the nursery soil with Formaldehyde (0.5%).
- Spray Agrimycin-100 or Streptocycline (50ppm) at transplanting, curd formation and pod formation.
- Crop rotation for 2-3 yrs with non cruciferous crop.
- Application of bleaching powder at 10.0 to 12.5 kg/ha controls the disease.
- Resistant Varieties: Cabbage: Cabaret, Defender, Gladiator, Pusa Muktha
- Cauliflower: Pusa ice, Pusa snow, ball-K-I-F, Sel-12

7. BACTERIAL SOFT ROT

Erwinia carotovora

Economic Importance

- Bacterial soft rot occurs on a wide range of vegetables all over the world and causes serious losses in the field, transit and storage.
- In northern India, especially in Kashmir, Kullu and Kalpa, where cabbage and cauliflower are grown for seed production, the rotting of heads and curds starts just after chilling and melting of snow.

Symptoms

- Small water soaked lesion appear on infected tissue, which enlarges rapidly in diameter and depth.
- The affected plants show a soft, slimy, foul smelling rot in a short time.
- The cauliflower curds and cabbage heads fail to throw flowering shoots followed by stump rot.
- Root crops exhibits symptoms on the lower parts of the stem which results in stunting, wilting and death of plants.
- The affected fields emit foul smell which is one of the main characteristic features of the disease.

Pathogen

- It is a gram negative rod shaped bacterium with 1 to 6 peritrichous flagella.

Disease Cycle

- The pathogen overwinters in infected fleshy organs in storage, on plant debris in the field, in the soil and in the pupae of insects.
- It only penetrates through damaged tissue, such as lesion caused by other pathogens or injury caused by extreme environmental conditions and cultural operations.

- Bacteria multiply on the substances released by the wounded cells and produce pectolytic and cellulolytic enzymes as a result of which cells plasmolyse, collapse and die.
- The collapsed cells are invaded by bacteria because their enzymes advance ahead of them.

Environmental Factors

- Infection is favoured by high humidity.
- Optimum temperature for disease development-25-30^{0}C.

Management

- Remove the infected surface leaves during sunny days.
- Disease tolerance varieties-Chinese cabbage, CC23, CCI, Majica and Arcadia.
- Spray Plantomycin or Streptocycline (500ppm) in combination with Copper oxychloride (0.3%) at fortnightly interval.

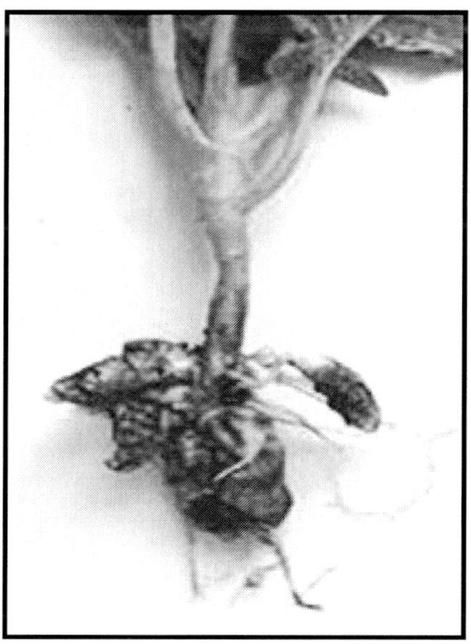

Club root

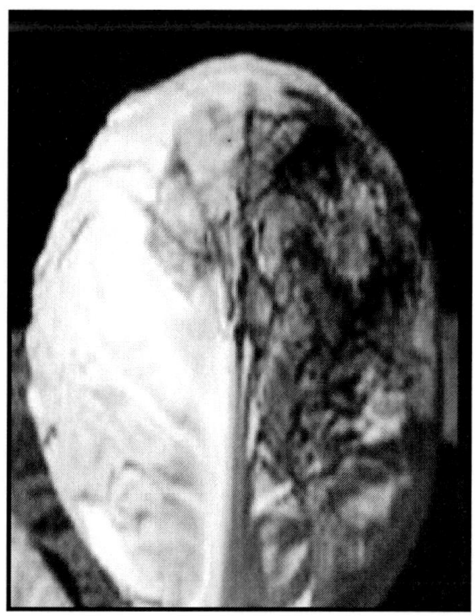

Downy mildew

White rust

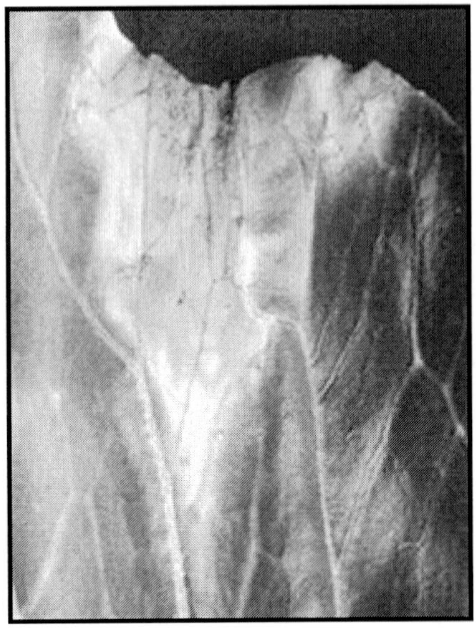

Black rot

Alternaria leaf spot

CHAPTER - 20

Diseases of Cucurbits and Their Management

SI. No	Major diseases	Causal organism
1.	Downy mildew	*Pseudoperonospora cubensis*
2.	Powdery mildew	*Erysiphe cichoracearum,*
3.	Anthracnose	*Colletotrichum orbiculare*
4.	Cercospora leaf spot	*Cercospora citrullina, C. melonis, C. lagenarium*
5.	Alternaria Blight	*Alternaria cucumerina*
6.	Bacterial wilt	*Erwinia tracheiphila*

1. DOWNY MILDEW

Pseudoperonospora cubensis

Economic Importance

- Downy mildew of cucurbits was first discovered in Cuba in 1968 but, at present it occurs in various part of the world.
- In India. It is fairly common in northern part where it causes severe loss during the later part of the rainy season.
- This is an important disease of cultivated cucurbits especially infects cucumber, muskmelon, ridge gourd, sponge gourd, bottle gourd, bitter gourd and snake gourd.

Symptoms

- Symptoms resembling mosaic viz, pale green areas separated by dark green areas appear on upper surface of leaf.

- During wet season, corresponding lower surface is covered with faint purplish fungal growth.The entire leaf dries up quickly.
- Fruits produced may not mature and have a poor taste.

Pathogen

- It is an obligate parasite.
- The mycelium is coenocytic and intercellular with small ovate or finger like haustoria. One to five sporangiophores arise through the stomata.
- Sporangia are grayish to olivaceous purple, ovoid to ellipsoidal, thin walled with a distal papilla.
- Zoospores are 10 – 13 micron meter, biflagellate.
- Oospores are not common.

Disease Cycle

- The pathogen survives on the diseased plant debris or wild cucurbits may overwinter as thick walled oospores.
- Sporangia produced on wild cucurbits, active mycelium on self sown or cultivated sponge gourd, and the oospores act as the source of primary infection.
- Sporangia and oospores germinate producing zoospores that are disseminated on to the host, germinate and cause infection.
- Secondary infection is caused by sporangia produced on primary infected host.
- Sporangia are disseminated by wind.
- The sporangia germinate on the leaf surface producing motile zoospores.
- Cucumber beetles are reported to carry the sporangia.

Environmental Conditions

- Free moisture on leaf and dew is important factors affecting infection and outbreak of disease.
- A minimum wetness period of three hour was found essential for initiating infection.

- The optimum temperature for infection is around 20^{0}C.
- A dew period of 2 hours is required for penetration and of 4 hours for haustoria formation.

Management

- Early sowing, weed host destruction, use of high nitrogen and phosphorus and low potassium fertilizers, and less moisture among the vines reduce the disease incidence.
- Destruction of cucurbitaceous weeds around field.
- Use of bed system with wide spacing with good drainage and air movement and exposure to sun help to check the disease development.
- Spray Metalaxyl (0.2%) or Chlorothalonil (0.2%) or Mancozeb (0.2%) at 15 days interval.

2. POWDERY MILDEW

Erysiphe cichoracearum (Sphaerotheca fuligena)

Economic Importance

- Powdery mildew of cucurbits has been reported from various cucurbitaceous crops including water melon, musk melon, pumpkin, cucumber, bottle guard, sponge guard, round melon, parwal and kundru.
- The disease has been recognized as a major disease both under field and green house conditions, the world over.
- In India, the disease was reported by Butler (1918) from UP and Bihar.
- Powdery mildew may result into considerable yield losses under dry weather conditions.

Symptoms

- The disease appear as superficial, powdery, grayish-white growth on upper leaf surfaces, petioles, and even main stems of infected plants.
- Affected areas turn yellow than brown and die.

- In dry seasons, powdery mildew can cause premature leaf drop and premature fruit ripening.

Pathogen

- The conidia measure 63.8 x 31.9 micron meter.
- The cleistothecia are dark, spherical with myceloid appendages which contain 10 – 15 asci.
- In each ascus, ascospores are two and are oval or sub cylindrical.
- However, cleistotheical stage is not very common.

Disease Cycle

- The fungus can overwinter in the form of mycelium and conidia on different cucurbitaceous hosts and this may serve as means of primary inoculums
- Wild cucurbits harbour the conidial stage of the fungus and release conidia for primary infection to the spring or summer sown cucurbits.
- Secondary infection occurs through conidia spread by wind, thrips and other insects, germinating giving rise to germ tube, and cause secondary infection on the host.
- When the pathogen approaches towards the end of the growing season, it may develop fruiting bodies, cleistothecia which may serve as perennating structures for the next growing season.

Favourable Conditions

- Heavy dew increases disease incidence.

Management

- Spray Calixin (0.1%) or Karathane (0.2%).

3. ANTHRACNOSE

Colletotrichum orbiculare (C. lagenarium)

Economic Importance

- Anthracnose of cucurbits is a relatively common disease of the foliage and fruits of water melon, musk melon, cucumber, bottle gourd, round melon, snake gourd etc under humid weather conditions.
- In India, the disease is very destructive in Karnataka during the rainy season.
- Yield loss of marketable fruits upto 99.5% have been reported.

Symptoms

- Lesions can form on seedlings, leaves, petioles, stems and fruits of susceptible cucurbits.
- Sunken, elongated stem cankers are most prominent on muskmelon, though leaf and fruit lesions also occur.
- Large lesions girdle the stems and cause the vines to wilt.
- Stem cankers are less obvious on cucumbers, but leaf lesions are very distinct.
- Watermelon foliage affected by anthracnose appears scorched; sunken fruit lesions are easy to recognize.

Pathogen

- The conidia are one celled hyaline, oblong to ovate 4-6 by 13-19 mm .
- The telemorph *Glomerella lagenarium* is rarely found in nature.
- The pathogen produces black stromata bearing black setae and hyaline conidiophores on host surface.
- Conidia are produced successively by budding at the tips of conidiophores accumulating in pinkish mass.

Disease Cycle

- The pathogen is seed borne, can also survive in infected plant debris, volunteer plants.
- Conidia are disseminated by wind, rain, implements.
- Spores are splashed from leaf to leaf, and plant to plant, during irrigation or rain events.

- Several disease cycles can occur in a single growing season, resulting in defoliation of severely infected plants.
- Visible symptoms appear up to 72 h after infection.

Environmental Conditions

- Humid rainy water is essential for infection.
- Spore germination is optimum at 22-27^{0}C.
- Relative humidity-100% for 24h.
- Invasion takes place up to 72h after conidial deposition.

Management

- Deep ploughing of crop residue immediately after harvest.
- Seed treatment with Carbendazim 2g/kg of seed.
- Spray Mancozeb 2g or Carbendazim 1 g/litre of water.
- Crop rotation in which no cucurbits are grown for at least one year period.

4. CERCOSPORA LEAF SPOT

C. citrullina, C. melonis, C. lagenarium

Economic Importance

- *Cercospora* leaf spot is a common disease of most cucurbits grown in India.
- During certain years the fungus causes devastating damage resulting in heavy loss in fruit yield.

Symptoms

- Initially minute water soaked spots appear on leaves.
- Spots enlarge rapidly and becomes circular to irregular with pale brown, tan or white centers and purple to almost black margins.
- Spots coalesce to form large blotches.
- The leaf may dry and die presenting the leaf a scorched appearance.
- Stems and fruits are also attacked.

Pathogen

The following are the records of *Cercospora* spp. on different cucurbits :

Name of cucurbits	*Cercospora* spp.
Momordica charantia	*C. momordicae, C.citrullina*
Tricho santhes anguina	*C. trichosanthes var. anguinae,C.citrullina*
Luffa acutangula	*C annamalaiensis,*
Lagenaria siceraria	*C. lagenariae*
Cucumis sativus	*C. chidambarensis*

Disease Cycle

- The disease is known to perpetuate only through the conidial stage.
- It may survive in the host tissue and other plant debris during the off season and also perhaps through the collateral hosts.

Management

- Maintain good soil drianage and good aeration between vines.
- 2 - 3 protective sprays with Zineb or Mancozeb (0.2%).

5. ALTERNARIA BLIGHT

Alternaria cucumerina

Economic Importance

- *Alternaria* leaf blight is also now becoming a very important disease of various cucurbits like watermelon, muskmelon and cucumber.
- The incidence varies from location to location and from one country to another.
- Under favorable weather condition, this disease can cause a considerable loss in yield of certain cucurbits.

Symptoms

- Lesions on leaves are the characteristic symptoms.
- Initially yellow brown, small but later turn into large brown necrotic areas with concentric zones.
- Muskmelons are more susceptible than other cucurbits to

Alternaria blight. Often muskmelon vines will be almost completely defoliated by this disease.

- The pathogen also may cause fruit injury.

Pathogen

- In water melon isolate, the conidia are 50.5 - 86.4 x 22.8 micron meter.
- Transverse septa vary from 1 to 9 and longitudinal septa range from 1 to 4.

Disease Cycle

- The fungus can survive as mycelium in refuse from diseased plant at least for one season and possibly two year in dry conditions.
- Fungus spores can survive in dry warm conditions for several months.
- Secondary infection occurs through air borne conidia .

Environmental Conditions

- Warm, wet weather favors development of disease.

Management

- Plant disease-free seed in fertile, well-drained soil.
- Practice crop rotation with unrelated crops.
- Destroy cucurbit weeds.
- Spray the crop with Mancozeb @ 2 g /lit.

6. BACTERIAL WILT

Erwinia tracheiphila

Economic Importance

- Bacterial wilt is an important disease problem in several countries including India.
- It often possesses threat to melon and cucumber production.

Symptoms

- Plant first appear dark green and then become necrotic as the wilt becomes irreversible.

- Wilt is more severe in the early season at rapid growth stage.
- In case of pumpkin and summer squash, plants may show dramatic wilt symptoms during the day when heat is more but partially recover by next morning.
- Leaves of affected plants become chlorotic and the margins become necrotic before the plant collapses and ultimately dies.

Pathogen

- The pathogen is a gram negative, rod shaped, motile bacterium with peritrichous flagella.

Disease Cycle

- Dispersal of bacteria within and across the field is through cucumber beetle vectors.
- These are mechanically transmitted by contact with contaminated beetle mouth parts when the vector feeds on stems and leaves.
- These bacteria generally reproduce and multiply at the wound sites and then enter the xylem vessels after which they move down the petioles and stem.
- Wilting may results due to vascular plugging by bacterial masses and formation of gums and resins subsequently.
- Spread of bacteria to healthy shoots takes place through the dissolved walls of adjacent xylem vessels.
- Subsequent infections result from infected plants serving as inoculums sources.

Environmental Factors

- Wilt is favoured by warm weather, adequate soil moisture, enough sunlight and balance nutrient concentration.
- Young and succulent plants are most vulnerable to wilt.

Management

- Wilted plant should be rouged out at early stage.
- Development of bacterial wilt resistant varieties
- Use systemic insecticides for control of cucumber beetles serving as vector.

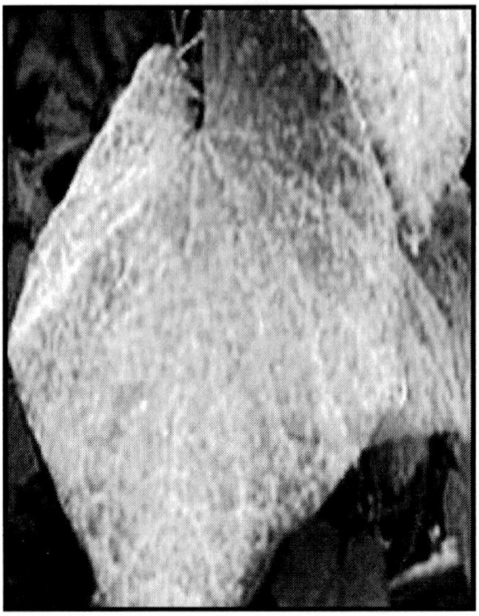

Powdery mildew

Downy mildew

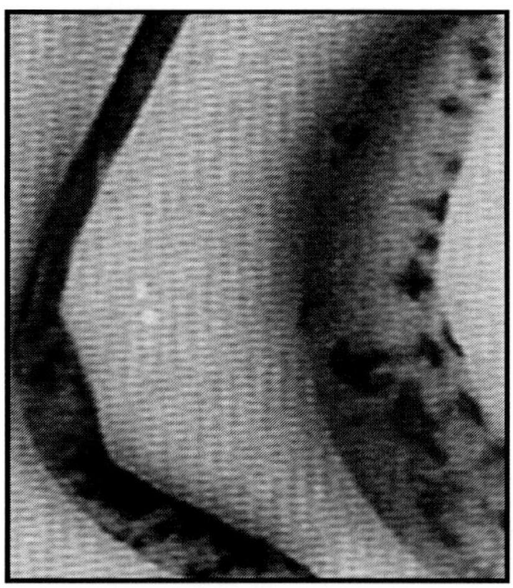

Anthracnose

Bacterial wilt

Fusarium wilt

Chapter - 21

Diseases of Onion and Garlic and Their Management

Sl.No	Major diseases	Causal organism
1.	Damping Off	*Pythium butleri* and *Fusarium oxysporum f.sp.cepae*
2.	Purple blotch	*Alternaria porri*
3.	Smut	*Urocystis cepulae*
4.	Downy mildew	*Peronospora destructor*
5.	Leaf blight (Blast)	*Botrytis* spp.
6.	Black mould	*Aspergillus niger*

1. DAMPING OFF

Pythium butleri and *Fusarium oxysporum f.sp.cepae*

Economic Importance

- This is a very common disease prevalent in onion nurseries but in garlic, the disease is not noticed.
- Damping off is one of the major constraints in the cultivation of kharif onion in northern parts of the country.

Symptoms

- Damping off occurs in two stages as described below:
- Pre emergence damping off: In this case the pathogen kills the radicle and plumule of the seed before to the emerging to the soil surface.

- Post emergence damping off: The pathogens attacks the collar region of seedlings at soil surface. The collar portion rots and ultimately the seedling collapses and dies.

Pathogen

- The fungus has a characteristic mycelium which is coenocytic, hyaline, freely branching and thick hyphae.
- It reproduces both asexually and sexually.
- The asexual reproduction is through the formation of sporangia.
- The zoospores germinate to produce a germ tube which gives rise to a characteristic coenocytic mycelium.
- The sexual reproduction of the fungus is characterized by the formation of oogonium, antheridium and oospore.
- The oospores may germinate when conditions are favourable, giving rise to germ tube and mycelium.

Disease Cycle

- The fungus is mainly soil borne.
- The oospores persist in soil, resisting adverse conditions.
- When there is sufficient moisture, they germinate and produce the mycelium which infect the host plant.
- The mycelia stage of the fungus is capable of infection the host plant and multiplying very rapidly.
- The zoospores formed in the vesicles of sporangia are commonly released in soil water to swim about and spread from place to place.
- When sexual reproduction starts, resulting in the formation of oospores which help to survive adverse environmental conditions.
- Thus, the fungus is capable of living for many years in soil, completing its life cycle both saprophytically and as a facultative parasite.

Management

- Used raised seed bed.

- Provide light, but frequent irrigation.
- Provide better drainage system.
- Drench the soil with Thiram or Captan @ 5g/m^2.
- Seed treatment with fungal culture *Trichoderma viride* (4 g/kg of seed) or Captan (2.5 g/kg of seed).
- Spray 0.2% Metalaxyl when there is cloudy weather.

2. PURPLE BLOTCH

Alternaria porri

Economic Importance

- This is an important and wide spread disease and is prevalent in almost all the onion and garlic growing areas of the world.
- In India, it appears in severe form, causing 20-50% damage to the onion seed crop.

Symptoms

- The characteristic features of the disease are small, whitish, sunken lesions with purplish centre which rapidly enlarge and eventually girdle the leaf or seed stem.
- Usually the affected leaf or stem falls down and dies within 3 or 4 weeks.
- In garlic the disease appears on leaves with the similar symptoms.

Pathogen

- Mycelium is branched, coloured and septate.
- Conidiophores arise singly or in groups.
- They are straight or flexuous, sometimes geniculate.

Disease Cycle

- Primary Infection: Plant debris and seed bulb.
- Secondary Infection: Air-borne conidia.

Favourable Conditions

- Warm humid weather with rains or heavy dew.
- Optimum temperature 22^{0}C.
- Relative humidity-90%.

Management

- Disease free bulb should be selected for planting.
- Seeds should be treated with Thiram @ 4 g/kg seed.
- The field should be well drained.
- Three foliar sprayings with Copper oxychloride (0.25%) or Chlorothalonil (0.2%) or Zineb (0.2%) or Mancozeb (0.2%).

3. SMUT

Urocystis cepulae

Economic Importance

- This is a dreaded disease of onion and garlic prevalent all over the world including India but in a localized manner.
- In India, the smut disease was noticed at Maysore.
- The pathogen causes considerable losses at the seedling stage.

Symptoms

- Pathogen attacks cotyledons of young plants soon after their emergence causing dark, elongated eruptive spots.
- On older leaves, the lesions may extend from base to the tip and appear as blisters.
- Lesions develop into thickened areas of several mm in size.
- Lesions burst open releasing masses of black smut spores.
- Severely affected plants killed within 3-4 weeks of emergence.
- Surviving plants are stunted with stout, brittle, distorted leaves bearing lesions throughout their length.
- Numerous blisters appear on leaves and bulb scales of mature plants which rupture to expose masses of black powdery spores.

Disease Cycle

- The spores are carried on the seed and bulb and are perpetuated from season to season in the soil.
- Secondary infection occurs through wind blown soil and surface drianage water, onion bulbs and onion transplants.

Favourable Conditions

- Optimum temperature 10-20^0 C.
- Plant is susceptible for 2-3 weeks from the date of sowing.
- Both onion and garlic are susceptible to this disease for a very short period i.e between germination and emergence.

Management

- Drenching of soil with 40% formaldehyde , 750 ml in 100 liters of water, at the rate of 5 litres to every 52 m of furrow, before planting.
- Seed treatment with Thiram or Captan @ 4g/kg seed. The bulbs may be dipped in Thiram solution (0.25%).
- Adequate control of both onion maggot and smut was obtained by the combination of an Organophosphate insecticide and Thiram.

4. DOWNY MILDEW

Peronospora destructor

Economic Imporatnce

- Downy mildew is another serious disease prevalent in all the subtropical and temperate regions of the world.
- The disease is reported on both onion and garlic.
- In India, the disease was first reported by Mir(1977) from hilly tracts of Jammu & Kashmir.
- The disease can rapidly destroy the foliage and thus prevent the formation, ripening, and storage of bulbs.

Symptoms

- A violet furry growth on the surface of leaves or seed stem is the main symptom of the disease.
- In onion and garlic crop where the symptoms appear on leaves, the affected plant are dwarf, stunted and become pale green in color.
- In severe cases bulb development is checked and bulbs do not attain full growth and are often soft and immature.
- In seed crop, the disease starts form mid portion and later spreads on the whole flower stalk.
- Flower stalk first become pale green on one side and later on violet colored mildew growth appears.
- As the umbel becomes heavier due to seeds the weakened stems break down which affects the seed maturation adversely.

Pathogen

- The coenocytic mycelium is intercellular with filamentous haustoria.
- The sporangiophores are non septate, long and swollen at the base.
- Sporangia are pyriform to fusiform, attached to the sterigmata by their pointed end. These sporangia germinate by one or two germ tubes.
- Oogonia are formed in the intercellular spaces.

Disease Cycle

- The main sources of perennation are the diseased bulbs used for propagating the crop in many areas and oospores present in diseased crop residues.
- When favorable environmental conditions occur, the over wintering fungal mycelium in systemically infected plants produces spores.
- After dissemination through the air, these spores infect the leaves of onion plants in commercial fields.

- The mycelium of downy mildew in leaves of infected onion plants in commercial bulb production fields produces a new crop of spores called conidia in cycles of approximately 11–15 days.
- As the upper portions of a leaf are killed, the fungus infects the next lower part of the leaf in each successive cycle of spore formation.
- Such cycles can be repeated several times until the leaf may be completely killed.
- These repeated cycles of spore formation can result in severe and continued epidemics of downey mildew if disease favorable environmental conditions persist.

Environmental Conditions

- Spores are formed at night when high humidity and an optimal temperature of 13°C occur.
- The spores mature early in the morning and are disseminated during the day.
- Germination occurs in free water with an optimal range of 7–16°C temperature.
- Rain is not needed for infection if heavy dews occur continuously during the night and morning hours.

Management

- Three spraying with Mancozeb (0.2%) is effective.
- Spraying should be started 20 days after transplanting and repeated at 15 days interval.

4. LEAF BLIGHT (BLAST)

Botrytis cinerae, B. squamosa and *B byssiodea*

Economic Importance

- The disease is common in temperate areas of the world viz. Netherlands and South-East Florida.
- The disease was first reported on onion in India by Gupta and Srivastava.

Symptoms

- The characteristic symptoms of the disease are small, superficial lesions developed on the leaves.
- Young spotted leaves start drying from tip and ultimately weak plants die.

Pathogen

- Botrytis is characterized by its conidiophores which present on grape bunch.
- The conidiophores are tall, erect and branches irregularly or dichotomously. They are dark and septate.
- The terminal cells swell to produce sporogenous ampullae.
- On each ampulla numerous conidia arise simultaneously on short lenticles.
- The conidia are hyaline or tinted, aseptate and globose to ovoid.

Disease Cycle

- The Botrytis leaf blight pathogen over winters as sclerotia.
- The sclerotia are formed on infected leaves and the necks and upper portions of bulbs before or after harvest.
- Sclerotia produce conidia (spores) that infect leaves on sprouted bulbs and onion plants in commercial fields.
- Sclerotia on leaf debris produce conidia and also ascospores (sexual spores) that infect leaves of onion plants.
- It is suspected that sclerotia in the soil and sclerotia on volunteer plants provide the primary source of inoculum for outbreaks of Botrytis leaf blight in commercial onion fields.
- Similar secondary cycles occur slightly later in seed fields. Conidia are blown from the seed fields and cull piles to commercial bulb production fields and the disease cycles continue.
- Leaves of onion plants in commercial fields can be infected by conidia that develop from sclerotia in the soil.

- These infected leaves also serve as secondary sources of inoculum once conidia are produced on the dead leaf tissue.
- Leaves of onion plants grown for commercial production are repeatedly infected, and these serve as the source of inoculum for infecting new leaves throughout the growing season.

Management

- Crop rotation and field sanitation.
- Avoid overlapping off onion crops.
- Bulb treatment with Captan /Thiram (0.25%).
- Spraying of Mancozeb or Chlorothalonil. Fungicides may be applied every 10 days for disease control.

5. BLACK MOULD

Aspergillus niger

Symptoms

- Infection usually is through neck tissues as foliage dies down at maturity.
- Infected bulbs are discolored black around the neck, and affected scales shrivel.
- Masses of powdery black spores generally are arranged as streaks along veins on and between outer dry scales. Infection may advance from the neck into the central fleshy scales.
- In advanced disease stages, the entire bulb surface turns black, and secondary bacterial soft rot may make the bulb soft and mushy.

Management

- Seeds should be treated with Thiram @ 4 g/kg seed.
- The field should be well drained.
- Three foliar sprayings with Copper oxychloride (0.25%) or Chlorothalonil (0.2%) or Zineb (0.2%) or Mancozeb (0.2%).
- Growers must follow crop rotation and harvested bulbs must be thoroughly cured to reduce potential storage losses.
- Soil drenching with Copper oxychloride (0.25%).

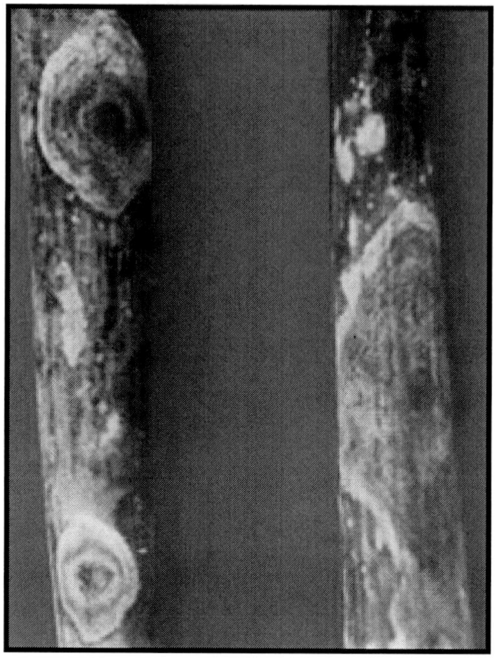

Purple blotch

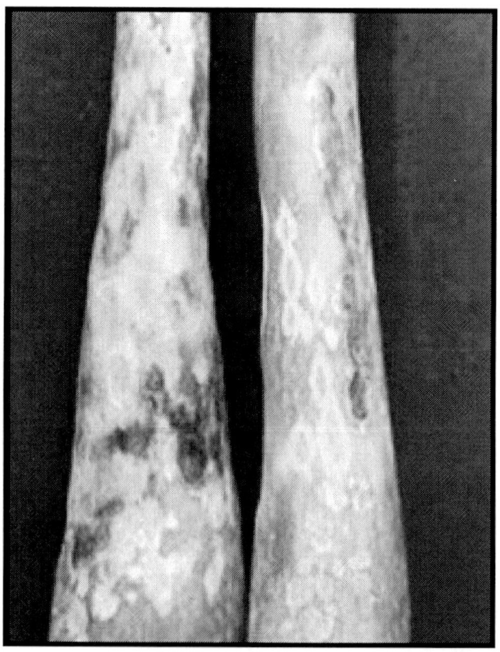

Downey mildew

Leaf Blight (Blast)

Smut

CHAPTER - 22

Diseases of Pea and Their Management

Sl. No	Major diseases	Causal organism
1.	Downy mildew	*Peronospora pisi*
2.	Powdery mildew	*Erysiphae polygoni*
3.	Rust	*Uromyces pisi*
4.	Wilt	*Fusarium oxysporum f. sp.pisi*

1. DOWNY MILDEW

Peronospora pisi

Economic Importance

- Downy mildew is present primarly during the early part of the growing season and is seldom of economic importance in most pea growing areas of India.
- It is evident commonly in Northern India and other areas where peas are grown under cool moist conditions.

Symptoms

- Symptoms are evident on leaves, stems and pods.
- First yellow to brown scattered patches appear on the upper surface of infected leaves and stipules and then grayish violet downy growth of the pathogen can be seen just opposite on the lower surface.

- Infected stems become distorted and show reduced growth resulting in subsequent stunting of the entire plant.
- Symptoms on pods are in the form of pale green, more or less elliptical to irregular, elongated lesions mostly confined along the dorsal sutures of the pods and in the later stage, they turn from light to dark brown in color.
- Seeds lying corresponding to the symptoms on the pods abort and become very much undersized.

Causal Organism

- The pathogen is an obligate parasite.
- Mycelium coenocytic, hyaline, profusely branched intercellular, finger shaped haustoria in side the host cells.
- The conidiophores are unbranched for about two third or more of their length and dichotomously branch at their tip.
- The branches are long, slender, pointed and are called sterigmata.
- A single conidium is produced at the tip of each sterigmata.
- The conidia are oval to elliptical and light brown in color.
- Oospores are produced at the end of the growing season.

Disease Cycle

- The pathogen perennates in the soil through oospores, which are brought therein along with infected plant debris.
- The oospores perennating in soil germinate by germ tube at the return of favourable conditions.
- The germ tube infects underground parts of young seedlings and the pathogen progresses upwardly systemically causing primary infection.
- The primary infection results in the production of conidia on infected leaves.
- The secondary infection occurs through wind blown conidia produced as result of primary infection .
- They are brought on the surface by the young pods, and cause secondary infection thereupon by means of each producing

germ tube, which penetrates through stomata and gives rise to local lesions.

Favourable Conditions

- Cool and moist weather conditions very much favour the disease incidence.

Management

- Field sanitation
- Practicing crop rotation for at least 2-3 years .
- Seed treatment with Metalaxyl (0.2%)

2. POWDERY MILDEW

Erysiphae polygoni

Economic Importance

- The disease is evident all over the world and causes great damage to the pea crop.
- It appear in epidemic form in the plains every ear when the pods are formed in January-February.
- Since the disease occurs more frequently and covers a large host surface, it proves much more destructive than the downy mildew diseases of the same crop.

Symptoms

- First symptoms observed on Leaves.
- White floury patches appear on both side of the leaves as well as tendrils, stems, pods, etc.
- As the plants become older, the symptoms almost cover the entire plant, become more or less grayish brown and the infected parts impart dirty appearance.
- In extreme severe conditions, the infected leaves are shed leaving stem devoid of them.

Pathogen

- The pathogen is an obligate parasite.
- Mycelium is septate, hyaline, profusely branched, and superficial forming a white web like coating over the surface of the infected part.
- Conidia are produced in chian on conidiophores.
- Mature conidium is elliptical or barrel shaped, hyaline, single celled.
- Cleistothecia, the fruiting bodies,are black, minute bodies provided with a number of mycelia appendages, and remain scattered in the mycelia web.
- Each cleistothecium contains usually 2-8 asci.
- Each ascus contains 3-8 ascospores which are hyaline, one celled and elliptical.

Disease Cycle

- The disease is soil borne as the pathogen perennates through cleistotheica, the fruiting bodies lying in soil on infected plant debris.
- After return of favourable conditions, the perennating cleistothecia get broken as a result of wall disintegration and release asci and ascospores.
- The ascospores are brought on to the lower most leaves near the soil by various agenicies, germinate by germ tube which cause primary infection on them.
- The conidia produced as a result of primary infection on host are wind blown, brought on to healthy leaves and other parts, germinate by germ tubes, result in secondary infection on them during the growing season.

Favourable Conditions

- Warm and dry weather conditions at 24-32^{0}C very much favour the disease incidence.

Management

- Field sanitation.
- Destruction of diseased plant debris.

- Crop rotation.
- Sulphur dusting (200 mesh) @25-30 kg/ha or spray with 0.2% Karathane effective in controlling disease.

3. RUST

Uromyces pisi

Economic Importance

- Rust disease is very wide spread and destructive to pea crop.
- It appears almost simultaneously with powdery mildew but develops very fast and causes greater loss.
- All most all the commercial varieties in cultivation viz, Banarasi Moti, Bonneville, Sweet, Little Marvel, Har Bonna, Harichhal show very heavy losses due to their high susceptibility to the disease.

Symptoms

- Numerous orange colored pustules having aecia in round or elongated clusters usually appear on the lower side of leaves.
- Later on, the symptoms manifest on all other aerial parts like stems, petioles, tendrils and even the pods.
- The orange colored pustules gradually turn brown (due to development of uredo-stage) during late in the growing season, and finally become almost black in color(due to development of teleuto-or telio-stage) when the crop moves towards maturity .

Pathogen

- Rust of pea is caused by two species of *Uromyces*, namely, *U.pisi* (pers.) Wint and *U fabae* (Pers.) de Bary.
- In India, however, the disease is most frequently caused by *Uromyces fabae,* which is an autoecious species as all its spore stages develop on the same host.
- The aecial stage appears first accompanied by pycnial stage.
- The pycnia are infrequent, hypophyllous, and rather inconspicuous.

- The aeciospores are rounded angular to elliptical, yellow with fine warts, and measure 14-22 µm in diameter.
- The uredial stage represented by uredopustules containing uredosori appears late in the season .
- Each uredosorus consists of numerous round to ovate, light brown, spiny uredospores, which measure 20-30x18-26 micron in size.
- The teleutostage represented by teleutopustules containing teleutosori appear most commonly on stems and petioles.
- Each teleutosorus contains subglobose, smooth, ovate or elliptical brown teleutospores.
- The teleutospores are stalked; the stalk being pale yellowish brown in color, persistant and measure upto 90 microns in length.

Disease Cycle

- The pathogen perennates *via.* teleutospores present on diseased plant debris or mixed with the seed as external contaminant.
- The primary infection on the newly growing crop is considered to be caused by the germination of teleutospores in the month of January.
- Aeciospores play a major role in the dissemination of the pathogen during the crop season resulting in secondary infection.
- The fungus may multiply freely on different host such as broad bean, lentil, sweet pea and lathyrus and pass from season to season without difficulty.

Favourable Conditions

- Persistence of low temperature and moist condition for few weeks in and around the infected field is thought to be favourable for severe disease incidence.
- The uredospore germinate usually at the temperature of 16-22.5^{0}C.

Management

- Destruction of diseased plant debris after harvest.
- Crop rotation
- Resistant varieties: PJ 22217, PJ 207708, E.C.42959, Perf.329,
- Foliar spray with Carbendazim (0.1%) or Benomyl (0.1%) at 15 days interval.

4. WILT

Fusarium oxysporum f. sp.pisi

Economic Importance

- Wilt is an important disease prevalent in those area where pea is cultivated in autumn.
- In India, the disease has been reported from Punjab, Delhi and West Bengal.
- The yield losses due to this disease have been recorded upto 13.9 to 95%.

Symptoms

- Dwarfing and yellowing of the foliage and there is downward curling of stipules and leaflets.
- At first the lower leaves and then the upper leaves show loss of turgidity.
- The entire plant wilts and the stem shrivels.
- The vascular tissues show a brown to orange or deep red discoloration which extends throughout the invaded root system.
- Fungal hyphae are restricted to the host xylem vessels and there is little or no rotting of the root cortex, although the pathogen invades from the soil through the cortex to reach the host vascular systems.
- The plant may die quickly or remain partly alive during the entire season.
- Pod formation is usually absent or seeds are absent in flattened pods.

Pathogen

- The disease is caused by *Fusarium oxysporum f. sp.pisi.*
- So far, 1,2,3,3A,4,5 and 6 races have been reported world over. But race 5 was most commonly occurring one.

Disease Cycle

- The fungus is soil borne, living saprophyticlly on soil organic matter.
- When the crop is harvested, the diseased roots are left over in the soil and the fungus survives on the host root for several years.
- It may produce chlamydospores, which are resting spores that can withstand adverse soil conditions.
- They become active when the next crop is sown.

Favourable Conditions

- Plant mortality was maximum in November at soil temperature of 23^0C and minimum in January at a maximum air temperature of 18^0C.

Management

- Late sowing (Between 1st and 15th November).
- Balanced application of fertilizers.
- Resistant varieties- JP501, JP-1140, JP Batri Brown 6 and JP-179.
- Seed dressing with Ceresan or Captan or Bavistin.

Powdery mildew

Downy mildew

Wilt

CHAPTER - 23

Diseases of Beans and Their Management

SI.No	Major diseases	Causal organism
1.	Anthracnose	*Colletotrichum lindemuthianum*
2.	Bean rust	*Uromyces appendiculatus*
3.	Yellow mosaic	Bean yellow mosaic virus, Mungbean yellow virus
4.	Common bean mosaic virus/ Green mosaic	Bean common mosaic virus *(ss RNA)*

1. ANTHRACNOSE

Colletotrichum lindemuthianum

Economic Importance

- Anthracnose occurs wherever beans are grown throughout the world but causes greater losses in the temperate zones than it does in the tropics.
- Serious losses from this disease have been reported.

Symptoms

- All the above ground parts are affected. However, the characteristic symptoms appear on pods.
- On cotyledons, spots are sunken dark brown or black with pink spore mass.
- Seedling infection results in collapse of seedling.

- Spots on leaves appear on lower side and are black. Later these may also appear on upper surface. When the infection is severe, the affected plants wither off.
- Black, sunken, circular spots of varying sizes appear on pods with bright red, yellow or orange margins.
- The centre of these spots later turns grey or pink due to sporulation of the pathogen. The border of these spots appear raised.

Pathogen

- Mycelium is branched, septate, hyaline at first and dark colored with age.
- Acervuli develop beneath the cuticle.
- Conidia are borne on short conidiophores.
- Setae are few, brown and septate. Conidia are one celled, hyaline and cylindrical with rounded ends or with one end slightly pointed.
- The fungus produces perithecia, which are 120-210 μm in diameter.

Disease Cycle

- The fungus overwinters in infected seeds and in plant debris.
- Conidia and /or dormant mycelia in the infected seeds germinate /become active and infect the young seedlings.
- It can remain viable for several years.

Favourable Conditions

- Relative humidity- 92% percent.
- Optimum temperature- 17^{0}C
- The fungus requires about 10 mm of rain to establish initial infection.

Management

- Use certified disease-free seed.
- Seed treatment with Carbendazim @ 2g/kg seed.

- Protect the crop by spraying 0.2% Benlate or Bavistin or Mancozeb (0.2%) at 15 days interval.

2. BEAN RUST

Economic Importance

- This disease is of common occurrence on different types of beans and cowpea in some pocket of the India.
- It may prove destructive through defoliation of the plant.
- The yield loss of dry bean due to this diseases varies from 2.7 percent to 36.7 percent.

Symptoms

- Symptoms mostly appear on leaves, rarely on stem and petiole, occasionally pods also show infection.
- The rust pustules appear on either sides but more common on lower surface.
- The rust pustules appear as minute , almost white, slightly raised spots.
- The Uredosori are minute, roundish, slightly raised and reddish brown coloured.
- These contain rust spores and appear in groups.
- Later in the season with the formation of teliospores, the sori turn dark brown or black.
- Diseased leaves may wither or fall off under severe infection.

Pathogen

- On Beans: *Uromyces phaseoli typica.*
- On cowpea and others: *U. phaseoli vignae.*
- On dry bean: *Uromyces appendiculatus.*
- Autoecious macrocyclic rust *i.e.,* produces all the stages on bean plant.
- Uredospores are echinulate, oval, 1-celled and yellowish brown in colour.

- Teliospores are dark brown coloured, elliptical and ovate, pedicillate, smooth walled, single celled with warty papillae at the top.
- Aeciospores are ellipsoid, 1-celled.

Disease Cycle

- The fungus survives through teliospores in crop debris in cooler regions and on collateral hosts.
- During the crop season the disease spreads through wind borne uredospores.
- Host range: French beans, green gram, black gram and cowpea.

Favourable Conditions

- The repeating spores (uredospores) germinate best at 15 to 24^0C.
- The teliospores in crop debris germinate best at 10-15^0C.
- Cloudy and humid days favours germination of spores and infection.

Management

- Adjust sowing dates to avoid severe infection of plants in the field.
- Protect the crop with Mancozeb or Zineb @ 0.2% or wettable sulphur @ 0.3%.

3. YELLOW MOSAIC

Bean Yellow Mosaic Virus, Mungbean yellow virus or Phaseolus virus-2 *(ss DNA virus)*

Symptoms

- Bright yellow patches appear on leaves. Yellow areas alternate with dark green areas of the leaf blade.
- Younger leaves show more severe mottling and chlorosis.
- Leaves completely turn yellow and gradually becomes necrotic.

- Plants are stunted and flower and pod set is reduced.
- Pod formation is reduced and if produced, they are deformed having shrivelled and undersized seeds.

Disease Cycle

- Primary Infection: Collateral hosts.
- Secondary Infection: Whitefly, *Bemisia tabaci.*
- Not transmitted by sap, seed and pollen.
- Host range: French bean, Soybean, Red gram, *Xanthium strumarum, Eclipta alba,* etc.

Management

- Remove collateral hosts and destroy.
- Use resistant varieties.
- Spray Metasystox @ 0.1% for vector control.

4. COMMON BEAN MOSAIC/GREEN MOSAIC

Bean common mosaic virus *(ss RNA)*

Economic Importance

- Bean common mosaic is one of the most commonly prevalent viral diseases of French bean, causing considerable losses in yield and quality.
- Yield losses amounting to 26 to 84 percent have been reported to be caused by the disease.

Symptoms

- Affects only beans (*Phaeolus vulgaris* & other *Phaseolus* species)
- Symptoms vary according to the variety of bean affected, time of infection and environmental conditions.
- Leaves show mosaic pattern *i.e.,* light green areas alternate with dark green areas.
- Diseased leaves become rough, and show blisters on the leaf lamina.

- Leaf size, petiole length and plant height reduced. Leaves curl downward.
- Diseased plants produce fewer pods which are smaller in size.
- Seeds become smaller, malformed and aborted.

Transmission

- Seed, sap, graft and aphid transmissible.
- Infected seeds are primary means of spread.
- *Aphis craccivora, Macrosiphum pisi, Aphis gossypi, A. medicaginis, Myzus persicae* and *Brevicoryne brassicae* transmit the virus in a non-persistent manner

Disease Cycle

- Infected seed, plants of susceptible French bean cultivars and weed hosts serve as natural sources of primary infection.
- Aphid vectors are responsible for secondary field spread of BCMV.
- Nymphs and apterous aphids are more efficient as compared to winged adults.

Management

- Use of disease free seed.
- Rouging of infected plants from the field or weed hosts from in and around the field helps to eradicate natural reservoir of the virus.
- Soil application of granular insecticides (Thimet @ 1.5kg a.i/ha) at sowing followed by foliar spray with either Monocrotophos (0.05%) or Dimethoate (0.05%) has been found effective to manage aphid population.

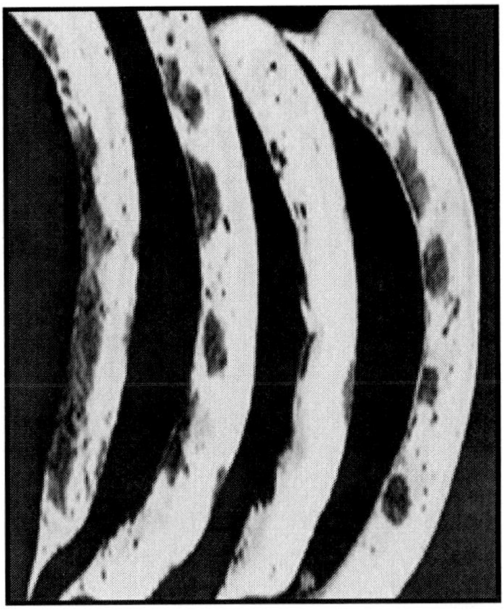

Anthracnose

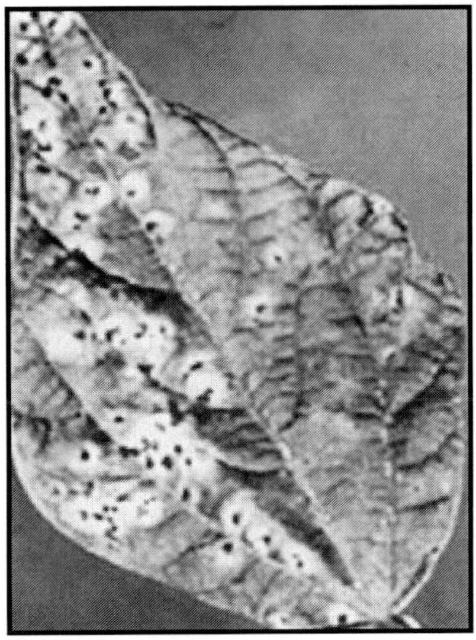

Bacterial blight

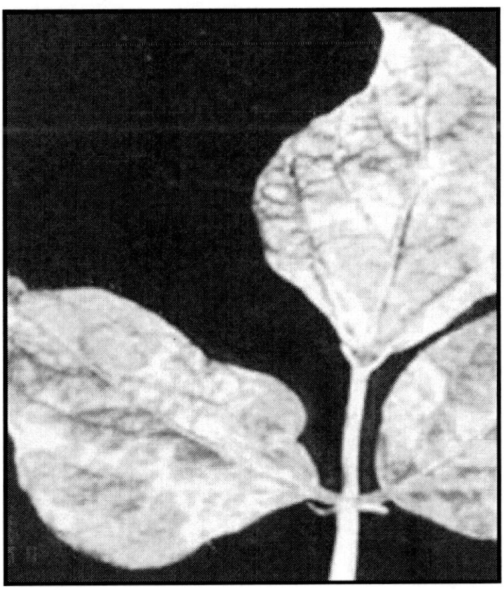

Common Mosaic

Chapter - 24

Diseases of Ginger and Their Management

Sl.No	Major diseases	Causal organism
1.	Rhizome rot/soft rot	*Pythium* sp.
2.	Leaf spot	*Phyllosticta zingiberi*
3.	Yellows disease	*Fusarium oxysporum f.* sp. *zingiberi*
4.	Banded leaf blight	*Rhizoctonia solani*
5.	Cercoseptoria leaf spot	*Cercoseptoria zingiberis*
6.	Bacterial wilt	*Ralstonia solanacearum*

1. RHIZOME ROT/SOFT ROT

Pythium sp.

Economic Importance

- Rhizome rot is the most serious disease of ginger.
- In India, the disease has been reported from almost all states like Rajasthan, Kerala, Himachal Pradesh, Orissa, Maharashtra, Tamil Nadu, Andhra Pradesh, Sikkim.
- The disease reduces potential yield to a greater extent in field, storage and market and may cause losses of more than fifty percent.

Symptoms

- The infection starts at the collar region of the pseudostems and progresses upwards as well as downwards.

- The collar region of the affected pseudostem becomes water soaked and the rotting spreads to the rhizome resulting in soft rot.
- At a later stage root infection is also noticed.
- Foliar symptoms appear as light yellowing of the tips of lower leaves which gradually spreads to the leaf blades.
- In early stages of the disease, the middle portion of the leaves remain green while the margins become yellow.
- The yellowing spreads to all leaves of the plant from the lower region upwards and is followed by drooping, withering and drying of pseudostem.

Pathogen

- *Pythium* sp. and *Fusarium solani*, either together or separately were found associated with rhizome rot.

Disease Cycle

- There are two ways in which the disease is carried over and perpetuated, firstly through the diseased rhizomes and secondly through oospores in soil.
- The infected plant debris remaining in the field forms important source of primary infection.
- The spread of the disease is typical as soil borne diseases because of fairly heavy and well distributed showers during the crop season from June to October.

Favourable Conditions

- A warm and humid climate predisposes the plant to infection at sprouting stage because of its tender and succulent tissues.
- Optimum temperature for growth and sporulation of *P.aphanidermatum* is about 34^0C.

Management

- Drench the field with Copper oxychloride @ 0.3% or Metalaxyl @ 0.2%.
- Application of *Trichoderma harzianum* along with neem cake @ 1kg/bed helps in preventing the disease.

- Collect the seed from disease free plots.
- Rhizome treatment with Metalaxyl @ 0.2%.
- Crop rotation for 3-5 years with non-host crops.
- Avoid water stagnation.

2. LEAF SPOT

Phyllosticta zingiberi

Economic Importance

- After soft rot, *Phyllosticta* leaf spot is the second most damaging disease.
- Leaf spot is reported to occur in different Indian states like Tamil Nadu, H.P and Kerala .
- This disease is prevalent from July to October.

Symptoms

- Initially disease appears as small, yellowish oval to elongated spots on the leaves measuring 0.5 to 1 mm.
- The spots enlarge; turn to white and papery at the centre with dark brown margin surrounded by a yellow halo.
- Spots are usually isolated but under humid conditions they coalesce with one another developing towards margin forming a big lesion.
- Black minute pycnidia develop in the white papery centre of the spots which is conspicuous in the later stages of the disease.

Disease Cycle

- The infected leaves in the debris of an infected field or seed serve as primary inoculum.
- The secondary infection occurs through conidia spread by splashing rain.
- The plants at 3-4 leaf stage are the most susceptible.

Favourable Conditions

- The disease appear towards the end of the June.

- In July, when the number of rainy days and total rainfall increase, the disease aggravates and spread fast.

Management

- Growing zinger in partial shade reduces the disease incidence.
- Spray Copper oxychloride (0.25%) or Mancozeb (0.2%) during monsoon.

3. YELLOWS DISEASE

Fusarium oxysporum f. sp. *zingiberi*

Economic Importance

- Ginger yellows, a serious stem rot, was first described from India by Haware and Joshi.
- This is a disease of relatively dry area.

Symptoms

- Yellowing of leaves starting from bottom to top.
- Drying of leaves, plants look like straw.
- Tillers do not fall to the ground.
- Shriveling/shrinking of rhizomes.

Pathogen

- The disease is caused by *Fusarium oxysporum f.* sp. *zingiberi* Trujillo.

Disease Cycle

- The disease is seed and soil borne.
- The seasonal carry over of the fungus inoculums takes place through infected rhizomes and soil.
- The fungus survives in soil as chlamydospores which may remain viable for many years in the field .
- The secondary spread of the disease can also take place through irrigation water and by mechanical means.

Favourable Conditions

- Maximum disease incidence occurred when soil temperature ranged from 24-25^{0}C and soil moisture from 25-30 percent.

Management

- Soak the seed rhizomes in a suspension of Carbendazim @ 1g/ litre before sowing.
- Two soil drenching with Mancozeb 2g/litre at 2 months.

4. BANDED LEAF BLIGHT

Rhizoctonia solani

Economic Importance

- The disease is prevalent in high rainfall zones (above 2000 mm) .
- It occurs from July and continues till the end of September.

Symptoms

- Distinct irregular bands of white papery patches appear on leaves.
- Shredding of leaves occur giving a ragged appearance to plant.
- At the end, plant becomes naked with only petioles left out.

Management

- Grow ginger under partial shade of dhaincha.
- Spray Carbendazim @ 1g/litre or Validamycin @ 2 ml/litre , twice at 15 days interval.

5. CERCOSEPTORIA LEAF SPOT

Cercoseptoria zingiberis

- The disease is important only at the higher altitude, i.e. 800m and above, in the hilly districts of Assam.
- It is severe also in the medium altitude of Meghalaya and Nagaland.

Symptoms

- Elongated yellowish brown lesions occur, running parallel to the leaf veins.
- The spots are white with dark brown margins.
- Black dot like structures are seen on the white portion of the spot.

Management

- Spray Bordeaux mixture @1% or Copper oxychloride @ 2.5g/lit twice at 15 days interval.

6. BACTERIAL WILT

Ralstonia solanacearum

Economic Importance

- Bacterial wilt and soft rot occur side by side in the same ginger field but on different plants.
- From severity point of view, bacterial wilt comes next to *Pythium* soft rot .
- Peak infection is in June and July.

Symptoms

- No yellowing of leaves.
- Mostly drooping and downward rolling of leaves is seen.
- Neck region is not soft.
- Stem cannot be pulled out easily.
- Most of the affected tillers remain in standing position.
- Leaves turn pale green.
- Milky ooze of bacteria is very clear from cut end of tiller.

Management

- Seed rhizomes should be collected from disease free area.
- Treat the seed with Copper oxychloride @ 3g/l at the time of sowing. Dip the rhizomes in the above suspension for 45 min. Ten litres of suspension is required for dipping 10 kg rhizomes,

at a time.The same fungicide is used for treating several batches of rhizomes.

- Two soil drenching with Copper oxychloride @ 3g/lit at two and three months after planting.if required, one more drenching should be given at four months after planting.

Rhizome rot

Bacterial wilt

Leaf Spot

Diseases of Flower Crops

CHAPTER - 25

Diseases of Rose and Their Management

SI.No	Major diseases	Causal organism
1.	Powdery mildew	*Sphaerotheca pannosa var. rosae*
2.	Die-back	*Diplodia rosarum*
3.	Black spot	*Diplocarpan rosae*
4.	Rust	*Phragmidium mucronatum*

1. POWDERY MILDEW

Sphaerotheca pannosa var. rosae

Economic Importance

- It is one of the widely distributed disease of rose.
- Powdery mildew is prevalent during Oct – Jan in south India and Dec- Feb in North India.

Symptoms

- Grayish powdery fungal growth on the infected leaves.
- Rolling of leaf margins with the onset of sprouting season.
- Affected leaves become curled and distorted.
- Badly infected flower buds fail to open.
- Infected floral parts become discoloured, dwarfed and dried.

Pathogen

- Mycelium is white, septate, ectophytic and sends globose haustoria into the epidermal cells of the host.
- Conidiophores are short and erect. Conidia are one celled, oblong, minutely verrucose with many large fat globules .
- Cleistothecia are formed towards the end of the season on the leaves, petals, stems and thorns.
- Cleistothecia are with simple myceloid appendages.
- Each ascus contains eight ascospores.

Disease Cycle

- The fungus survives as mycelium in dormant buds and shoots or as cleistothecia.
- Secondary infection occurs through wind borne conidia.

Management

- Collection and burning of fallen leaves.
- Resistant varieties: Ashwini, Ambika, Angeles, American pride, Surabhi.
- Sprays with wettable Sulphur @ 0.3% or Carbendazim @ 0.1% or Karathane @ 0.1% at 15 days interval.

2. DIE-BACK

Diplodia rosarum

Economic Importance

- Dieback was first reported in India in 1961 from Delhi.
- Now, the disease occurs in all the rose growing areas.

Symptoms

- Infection starts from the pruned surface of the twigs.
- The infected portion become brown to dark brown or black.
- Infected twig dries from tip downwards.

- Infection spreads from twig to main stem and roots and finally whole plant is killed.
- Older plants and neglected bushes are more frequently attacked.

Pathogen

- The fungus produces round, black pycnidia which bear spores.
- The pycnidiospores are dark coloured and two celled.
- Perithecia are immersed in the host tissue and are surrounded by a pseudostroma.
- Ascospores are ellipsoidal or fusoid, hyaline, two celled with the septum in or near the middle.

Disease Cycle

- The fungus survives as pycnidia present in dead twigs.
- The conidia spread primarily through irrigation water.

Management

- Pruning should be done so that lesions on the young shoots will be eliminated.
- Apply Chaubatia paste in the pruned area. Spray with Copper Oxy Chloride (0.2%) Difolatan (0.2%) Chlorothalonil (0.2%) Mancozeb (0.2%).
- Resistant varieties: Blue moon, Red gold, Summer queen, etc.

3. BLACK SPOT

Diplocarpan rosae (Anamorph: Marssonina rosae)

Economic Importance

- Black spot of rose is a serious problem in chill and cold climate of temperate regions. The disease causes marked reduction in the size and number of flowers.

Symptoms

- Dark brown tar coloured spots with fringed borders.
- Spots coalesce forming large patches.
- Infected leaves turn brown and defoliate.
- Fungus may also attack stems and flowers of rose bushes.
- On stems, infected areas are blackened with blistered appearance dotted with pustules.

Pathogen

- The vegetative body of the fungus consists of two parts *viz.*, the subcuticular mycelium and the internal mycelium.
- The fungus produces acervuli on the central part of the tar spots as blister like projections.
- Asci are discoid, sub epidermal and erumpent. Stroma is thin.
- Conidiophores are hyaline short and cylindrical.
- Conidia are hyaline, two celled, fusiform or allantoid to obclavate, upper end round, base narrow, guttulate.

Disease Cycle

- The fungal spores are spread primarily by splashing rain or water.
- Germination of the spores and infection occur when free water remains on the leaf surface for a period of 6 hours or longer.
- Leaf spots develop within 5 to 10 days.

Management

- Affected parts should be collected and destroyed.
- Space roses far enough apart for good air circulation.
- Spray Tridemorph @ 0.025% or Benomyl @ 0.1% at weekly intervals starting with the sprouting of the plants till new foliage appears.
- Avoid overhead watering and keep foliage as dry as possible.
- Resistant varieties: Bebe bune, Coronado, Grand opera, Sphinx. Radiance - (escape infection due to waxy surface).

4. RUST

Phragmidium mucronatum

Economic Importance

- Rose rust is restricted to higher altitudes.
- It occurs in Jammu and Kashmir, Himachal Pradesh, Punjab, Tamil Nadu and Uttar Pradesh.
- Rust disease outbreak was reported from Udaipur district of Rajasthan in the variety Chaiti Gulab.

Symptoms

- Damage to lemon yellow pustules appear on lower surface of the leaves and stems. Then the color changes to blackish red.
- The affected leaves turn yellow deformed and fall prematurely.
- Die back symptom also appear due to weakening of the plant.

Pathogen

- Aecidiospores are verrucose and orange yellow.
- They are surrounded in the aecidium by club shaped paraphyses.
- Uredospores are ellipsoid or ovate, echinulate and orange yellow.
- The uredospores are borne on short pedicels and are surrounded by paraphyses.
- Teleutospores are dark coloured, cylindrical, 6- 8 celled with a pointed papilla.

Management

- Collection and burning of fallen leaves.
- Spray with Carboxin (0.1%) or Wettable Sulphur (0.25%) or Captan (0.2%).

Powdery mildew

Die back

Rust

Black spot

CHAPTER - 26

Diseases of Chrysanthemum and Their Management

SI.No	Major diseases	Causal organism
1.	Wilt	*Fusarium oxysporum* f.sp. *chrysanthemi*
2.	Rust	*Puccinia chrysanthemi*
3.	Septoria leaf spot	*Sepotria chrysanthemella*
4.	Powdery mildew	*Oidium chrysanthemi*
5.	Chrysanthemum stunt	Viroid

1. WILT

Fusarium oxysporum f.sp. *chrysanthemi*

Symptoms

- Initial symptoms are in the form of yellowing and browning of leaves.
- Affected leaves die from the base of the plant upward.
- Infected plants are stunted and often fail to produce flower.
- Wilting may cause rotting of root or the base of the stem.

Pathogen

- Mycelium is white. Microconidia are formed singly, hyaline and cylindrical.
- Macro conidia are cylindrical to falcate.
- Chlamydospores are globose to oval and rough walled.

Disease Cycle

- The fungus is soil borne.
- The disease spreads through cuttings.
- Infection is always through injured roots.
- Deep wounds to expose the xylem help in easy infection.
- After entry into the root, the fungus multiply extensively in vascular tissues, interrupting the transport of nutrients to the above ground plant parts, thus stunting the plant growth.
- Continued dislocation of flow of plant nutrients leads to plant wilt.

Management

- Drenching the soil with Carbendazim (0.1%) is effective.
- Before planting dipping the rooted cuttings in a solution of Thiram @1.5g/litre of water.
- Since the disease spreads mostly through cuttings, it is important to use disease free planting material.
- Disease can further be minimized by following strict sanitation; periodical monitoring; crop rotation and roguing of infected plants.

2. RUST

Puccinia chrysanthemi

Symptoms

- The disease symptoms are in the form of brown blister-like swellings, which appear on the undersides of leaves.
- These burst open releasing masses of brown, powdery spores.
- Severely infected plants become very weak and fail to bloom properly.

Management

- Early removal of infected leaves/plants helps to prevent the further spread of the disease.

- Spray the plants with Propiconazole (1%) is effective in controlling the disease.

3. SEPTORIA LEAF SPOT

Sepotria chrysanthemella

Symptoms

- Blackish brown circular to irregular spots on leaves which later form large patches.
- The large patches covers major portion of the leaf.
- Patches are surrounded by chlorotic halo
- Later the centers turn greyish and the leaves remain small and curl
- Dead leaves hang on stem for some time

Pathogen

- Pycnidia are numerous, amphigenous, sub epidermal, globose or lens shaped.
- Conidia are hyaline, filiform, straight or flexuous often curved or worm like.

Disease Cycle

- Primary Infection: Infected plant debris in soil.
- Secondary Infection: Wind borne conidia.

Management

- Collect and burn diseased plant material.
- Spray Carbendazim @ 0.1%.
- Tolerant varieties: Alpana, Sarad and Flirt.

4. POWDERY MILDEW

Oidium chrysanthemi

Symptoms

- Infection is more severe in older plants under humid conditions.

- The growth of the fungus on the leaves appears as powdery coating. Infected leaves turn yellow and dry out.
- Infected plants remains stunted and fail to flower.

Pathogen

- Conidiophores are septate.
- The cleistothecia are sharp and globose.

Disease Cycle

- Primary Infection: Mycelium in infected plant debris.
- Secondary Infection: Wind borne conidia.

Management

- Disease can be effectively controlled with Sulphur Wettable (0.3%) or Karathane (0.1%).
- Good ventilation and proper spacing for free circulation of air is recommended.

5. CHRYSANTHEMUM STUNT

Viroid

Symptoms

- Small plants with paler leaf and flowers.
- Some flowers appear bleached and are inferior in quality.
- Flowers from diseased plants open 7 to 10 days earlier than healthy ones.
- Axillary buds grow prematurely and produce number of branches and stolons.
- Plants are stunted in appearance.
- White specks and yellow blotches appear on leaves of some varieties.
- Cuttings from diseased plants root poorly.

Disease Cycle

- **Primary Infection:** Viroid survives in infected plants
- **Secondary Infection :** Spread through sap and knives or tools used during pruning or pinching plants, taking cuttings and cutting the flowers.

Management

- Use certified Viroid free stocks.
- Remove and destroy infected plants.

Wilt

Rust

Septoria leaf spot

CHAPTER - 27

Diseases of Jasmine and Their Management

SI. No	Major diseases	Causal organism
1.	Rust	*Uromyces hobsonii*
2.	Cercospora leaf spot	*Cercospora jasminicola*
3.	Collar rot and Root rot	*Sclerotium rolfsii*
4.	Phyllody	*Phytoplasma*

1. RUST

Uromyces hobsonii

Symptoms

- Most of the plant parts are affected.
- Blisters or tumors and other abnormalities may be produced on the infected plant parts.
- Infected portions become hypertrophied.
- Orange coloured pustules on both surfaces of leaf .
- Chlorosis & defoliation of leaves under severe conditions.
- Yellow or orange coloured cankers on stems and twigs.
- Splitting of bark of affected branches and subsequent death of branches.
- Infected flower buds are swollen and deformed.
- Flower production is highly reduced.

Pathogen

- Autoecious rust (Demicyclic rust)
- Pustules are initially yellow (aeciospores) and later turn dark (teliospores)
- Uredial stage is absent

Disease Cycle

- Primary Infection: Teliospores in infected plant.
- Secondary Infection: Wind borne aeciospores.

Favourable Conditions

- Rust initiates in June- July and persists till March.

Management

- Remove and destroy affected plants or plant parts to avoid further spread.
- Spray Mancozeb @ 0.2%.
- Dust Sulphur (20-25kg/ha) or spray Tridemorph (0.05%) or Triadimefon (0.05%).

2. CERCOSPORA LEAF SPOT

Cercospora jasminicola

Symptoms

- Circular to irregular, reddish brown spots with a darker border and whitish centers appear on both the leaf surfaces and other aerial parts of the plant.
- Severe infection results in defoliation and die back symptoms.
- Flowering is adversely affected.

Pathogen

- The fungal mycelium is present both inter- and intra cellularly.
- It produces clusters of dark and septate condiophores in the host tissue.

- The conidia are linear, whip like, hyaline, whip like , thin walled, septate and are borne singly.

Disease Cycle

- The primary infection originates with the fungus getting carried over on plant debris or by seed.
- The fungus spreads mostly by air borne conidia.

Management

- Collect and burn diseased plant debris.
- Spray Carbendazim @ 0.1% or Copper oxychloride @ 0.3%.
- Monthly sprays of Benomyl or Thiophanate methyl @ 0.1%.
- Pari Mullai is moderately resistant.

3. COLLAR ROT AND ROOT ROT

Sclerotium rolfsii

Symptoms

- Plants at all stages are infected.
- First the older leaves become yellow followed by younger leaves and finally death of the plant.
- In the root black discoloration can be seen.
- On the infected tissues and stem surface white strands of mycelia and mustard like sclerotia are seen.

Management

- Soil drenching with Copper oxychloride (0.25%).
- Heavy application of FYM with *Trichoderma viride.*

4. PHYLLODY

Phytoplasma

Symptoms

- Leaves become small malformed and bushy.

- In the place of flowers green leaf like malformed flowers are formed.

Mode of Spread

- The disease is transmitted by grafting and whitefly, *Dialeurodes kirkaldii.*

Management

- Selection of cuttings from healthy plants.
- Spraying insecticide to control the vector.

Cercospora leaf spot

Collar rot and Root rot

CHAPTER - 28

Diseases of Crossandra and Their Management

SI. No	Major diseases	Causal organism
1.	Wilt	*Fusarium solani*
2.	Stem rot	*Rhizoctonia solani*
3.	Alternaria leaf spot	*Alternaria amaranthi* var. *crossandrae*

1. WILT

Fusarium solani

Symptoms

- Usually observed one month after transplanting in patches.
- Leaves of infected plants become pale and droop.
- Leaf margins show pinkish brown discolouration.
- Discolouration spreads to midrib in 7 to 10 days.
- Stem portion gets shrivelled.
- Dark lesions noticed on the roots extending upto collar region which results in roting of cortical tissue.

Disease cycle

- **Primary Infection:** Chlamydospores survive in soil.
- **Secondary Infection:** Microconidia dispersed through irrigation water.

Favourable Conditions

- Incidence is more in presence of root lesion nematode, *Pratylenchus* sp., and *Helicotylenchus dihystera.*

Management

- Remove and destroy affected plants
- Soil application of Phorate @1g/plant on 10th day of transplanting for nematode control.
- Soil drench with Carbendazim @ 0.1% or Copper Oxy Chloride @ 0.25%. Repeat soil drenching after 3-4 weeks

2. STEM ROT

Rhizoctonia solani

Symptom

- Brown to black lesions develop on stem just above soil level and result in girdling of the stem.
- The lesions extend to the upper part of the stem and result in collapse of seedlings.
- The roots are also rotted.

Management

- Soil drenching with Fosetyl-Al

3. ALTERNARIA LEAF SPOT

Alternaria amaranthi var. *crossandrae*

Symptoms

- Infected leaves show small, circular or irregular yellow spots on the upper surface.
- They soon enlarge turn brown and develop dark brown concentric rings.
- Infected leaves become yellow and drop off prematurely.

Management

- Spraying with Benomyl (0.1%) or Mancozeb (0.2%) or Carbendazim (0.1%).

Colour Plates

Diseases of Mango

Anthracnose

Powdery mildew

Flower malformation

Sooty mould

Red rust

Diseases of Banana

Panama wilt

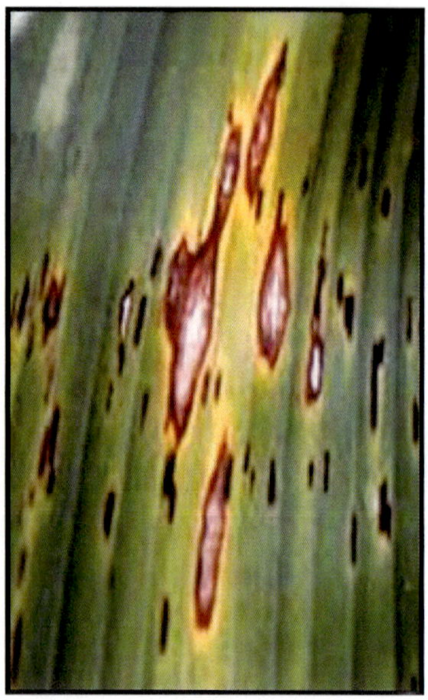

Sigatoka leaf spot

Anthracnose

Bacterial wilt

Bunchy top

Diseases of Grapes

Downy mildew

Powdery mildew

Anthracnose

Diseases of Citrus

Canker

Gummosis

Scab

Tristeza or quick decline

Greening

Diseases of Papaya

Collar rot

Powdery mildew

Anthracnose

Ring spot

Leaf curl

Diseases of Apple

Scab

Powdery mildew

Fire blight

Diseases of Pomegranate

Cercospora fruit spot

Leaf Spot or Blight:

Bacterial leaf spot

Diseases of Sapota

Sooty mould

Leaf spot

Flat Limb

Diseases of Coconut

Bud rot

Basal stem rot

Diseases of Coffee

Rust

Brown leaf spot

Anthracnose

Diseases of Tea

Blister blight

Red rust

Diseases of Betelvine

Leaf spot

Anthracnose

Bacterial blight

Diseases of Potato

Early blight

Late blight

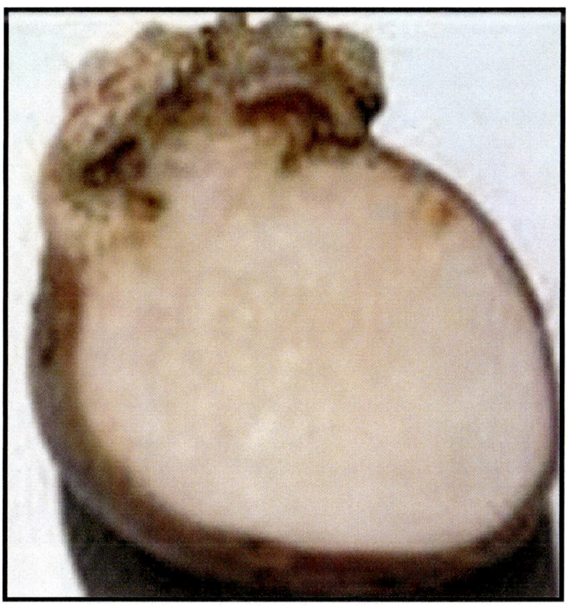

Wart

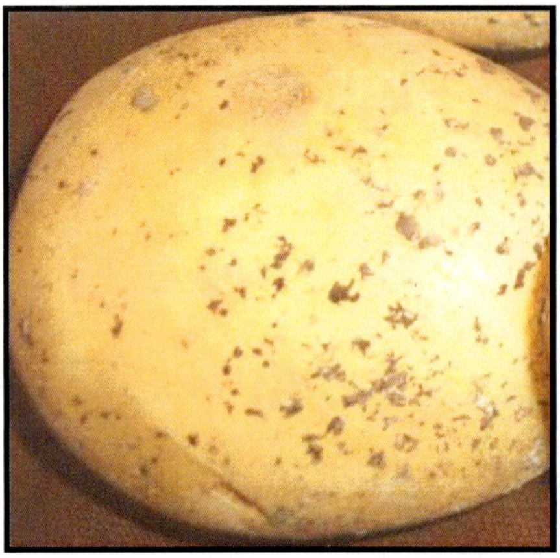

Black Scurf

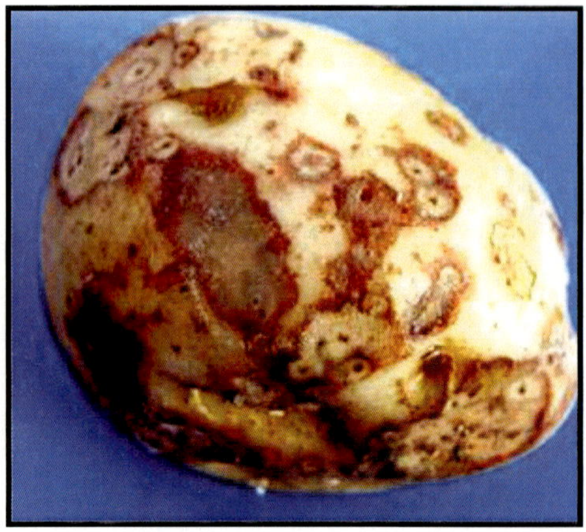

Black leg/Soft rot

Diseases of Tomato

Early blight

Late blight

Damping off

Fusarium Wilt

Mosaic

Leaf Curl

Bacterial Wilt

Diseases of Brinjal

Phomopsis blight

Little leaf of brinjal

Bacterial wilt

Collar rot

Diseases of Chilli

Damping off

Fruit rot

Powdery mildew

Cercospora leaf spot

Mosaic

Leaf Curl

Fusarium Wilt

Diseases of Bhendi

Yellow vein mosaic

Powdery mildew

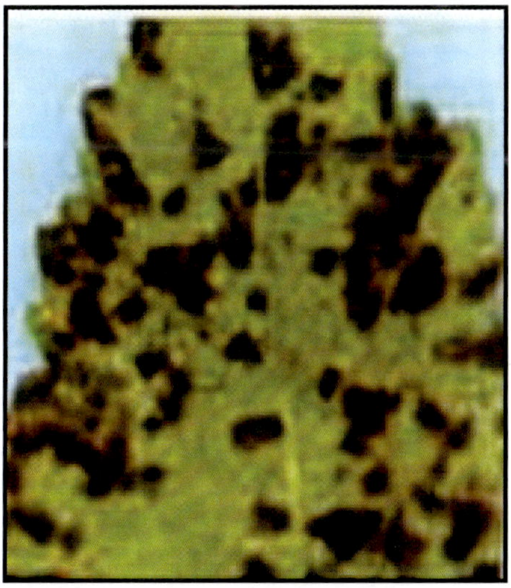

Cercospora leaf spot

Diseases of Crucifers

Club root

Downy mildew

White rust

Black rot

Alternaria leaf spot

Diseases of Cucurbits

Powdery mildew

Downy mildew

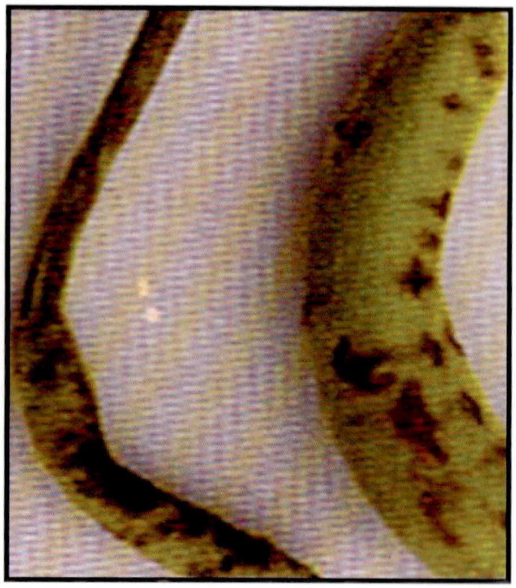

Anthracnose

Bacterial wilt

Fusarium wilt

Diseases of Onion and Garlic

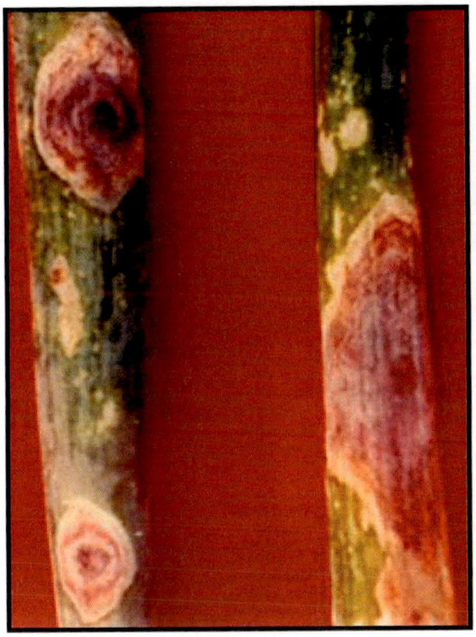

Purple blotch

Downey mildew

Leaf Blight (Blast)

Smut

Diseases of Peas

Powdery mildew

Downy mildew

Wilt

Diseases of Beans

Anthracnose

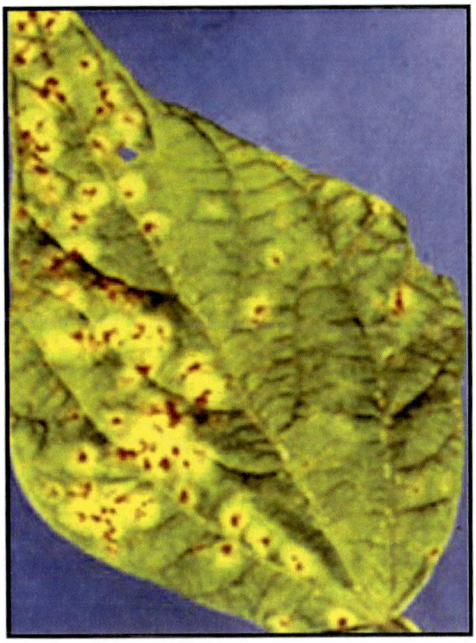

Bacterial blight

Common Mosaic

Diseases of Ginger

Rhizome rot

Bacterial wilt

Leaf Spot

Diseases of Rose

Powdery mildew

Die back

Rust

Black spot

Diseases of Chrysanthemum

Wilt

Rust

Septoria leaf spot

Diseases of Jasmine

Cercospora leaf spot

Collar rot and Root rot